BIOLOGICAL
ILLUSTRATION

Techniques and Exercises

JOHN C. DOWNEY
JAMES L. KELLY

BIOLOGICAL
ILLUSTRATION

Techniques and Exercises

Iowa State University Press
A M E S

Printed by The Iowa State University Press, Ames, Iowa 50010

First edition, 1982

International Standard Book Number: 0-8138-0201-6
Library of Congress Catalog Card Number: 82-83308

CONTENTS

PREFACE

The objectives of a short series of exercises in biological illustration are necessarily limited by time. However, significant progress may be made in several areas: increasing skill with pen and pencil; development of limited skills in pen-and-ink (line) illustration; and acquaintance with some common illustrating techniques and materials.

A course in biological illustration can sharpen perception; it can teach us to observe more closely. While truth and accuracy are essential in science, it is amazing how few scientists make any attempt to increase their skills in observation. Accuracy in appraising such things as size, shape, proportion, texture, tone, and the many details necessary to illustrate accurately a relatively simple biological object is a decided step beyond casual observation of nature. Biological illustration helps develop precision and exactness, both powerful tools of science.

Biological illustration also opens new avenues of communication for a science student. Its value in self-communication in the laboratory is obvious; the power of recall stimulated by a laboratory drawing is directly proportional to the accuracy and detail of the illustration. The growth of the graphic arts in the last decade has dramatically verified the value of drawing as an aid in communication. Relatively simple overhead projections or chalkboard illustrations may give a new visual impact to verbal concepts and ideas.

The objectives of this book are simple: (1) to introduce a series of exercises in a designed sequence moving from the simple and easy to the more involved and difficult; (2) to describe and develop students' skills in some of the basic illustration techniques and materials; (3) to help students develop self-critique methods necessary in mastering skills and interpreting published works; (4) to present some useful tips to use in completing the illustrations described in the exercises; (5) to build up students' confidence, motivation, and enthusiasm by using as illustrations drawings by other introductory students; and (6) to further assist students and teachers by supplying an annotated list of supplies and a list of references.

While it is our purpose to introduce the basic principles and techniques of pen-and-ink illustration, it should be recognized that this book does not cover the entire subject. However, to coin an illustrative phrase, "we had to draw the line somewhere."

Talent is a subjective concept, seemingly easier to recognize than to describe. It is a "weasel word" that sounds precise and meaningful, particularly to those who claim to have it. But you countless "untalented" souls who find comfort by announcing you weren't born with talent may still grow in the area of biological illustration, provided you have that all-important quality—willingness. How often have you heard someone belittle himself, saying, "I can't draw a straight line," as though this was the most significant value of the art world. Each of us has different qualities; while skill at drawing, or a steady hand, may help a person create an illustration, his skill will be of little help if he lacks the quality of careful observation, of attention to every detail of his subject; accuracy is the most sought-after virtue in biological illustration. When developing creativity, every stage of progress becomes a goal in itself, moving toward individual growth and enrichment.

It is hoped that this experience in biological illustration will be a step toward your individual growth and enrichment. Talent in biological illustrating, however modest, can be developed, building confidence and providing an additional means of communication for you in a variety of fields. But it requires your willingness to extend yourself, to find the upper limits of your talent. We hope it will be an enjoyable and rewarding route.

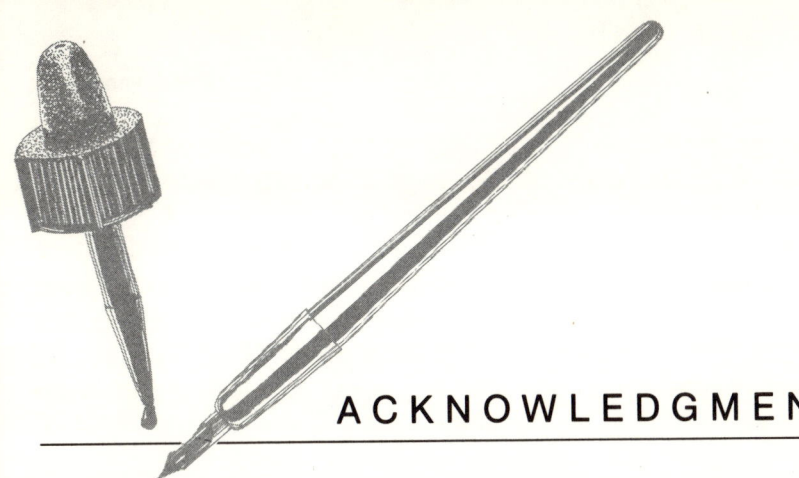

ACKNOWLEDGMENTS

We gratefully acknowledge the assistance of students who were challenged while they took a beginning course in biological illustrations to provide drawings for some mimeographed exercises originally provided for the course. These exercises and the resulting drawings provided the stimulus for James L. Kelly, who took the course while teaching biology at Malcolm Price Laboratory School, University of Northern Iowa, to suggest that a published edition would prove of some value to students and teachers alike.

We thank the following students for permission to use their illustrations cited by figure number in the text.

Keith Anliker . . . . .Figure 4.12
Christine D. Bradner . . . . Figures 4.16, 5.11, 5.13
Kathleen J. Briden. . . . . .Figure 5.1
Robert Conrad . . . . .Photograph, Figure 3.2
Pauline M. Drobney. . . . . .Figure 4.3 (part)
Sue Enghardt. . . . . Figures 4.9, 4.13, 4.14, 4.15, 4.17, 5.7
Andrew J. Franklin. . . . . Figures 5.4, 5.6
Robert S. Frenchick . . . . .Figure 4.20
Holly M. Hegenbarth. . . . . Figures 4.3 (part), 5.5, 5.14
David Heer . . . . .Figure 4.4
Dan Higgins. . . . . Figures 1.7, 1.8, 1.10, 1.14, 2.6B, 2.12,
2.13, 4.7, 4.8, 4.11, 4.22
Tim Ivers. . . . . Figures 1.15, 2.4, 2.9 (part), 6.4
James L. Kelly. . . . . .Figures 1.3, 1.4, 1.6, 2.1, 2.3, 2.7,
2.8, 2.10, 2.11, 3.I, 4.4, 4.18A, 4.18B, 6.1, 6.2, 6.3
Ken Kortage . . . . Figures 4.1B, 4.6

Kathleen M. Mayne . . . . Figures 1.1, 1.2, 1.5, 1.9, 1.11, 1.12
                                              1.13, 2.2, 2.14, 5.8, 5.10
Paula L. Navara . . . . .Figure 4.21
Marla R. Ohrt . . . . Figures 5.3, 5.9, 5.12
Craig J. Ones . . . . .Figure 4.10 (part)
Gary Simmons. . . . . Figures 4.10 (part), 5.2
Michael R. Smith. . . . . . . Figures 2.5, 2.6A, 3.3, 4.1A, 4.2, 4.5, 4.18C
Thomas L. Wagner. . . . . .Figure 4.20

BIOLOGICAL ILLUSTRATION

Techniques and Exercises

1

Using the Basic Instruments: Pen and Pencil_____

The pencil and pen are the most important tools between you and a biological illustration. Until they are mastered, handled with skill, a satisfactory drawing will not be possible. One sets about mastery of the pencil and pen by use—practice, practice, and more practice. Just as the typist does not develop speed and accuracy overnight, or a long-distance runner stamina by thinking about it, a biological illustrator must look to the fundamentals and exercise them repeatedly. The skill comes after the brain has decided the artistic objectives are worth the time and patience required to practice.

Exercise 1 and 2 consist of a series of exercises and drawings designed to familiarize you with using the pencil and pen in drawing simple lines and figures and to make you, the aspiring illustrator, aware of your ability. If you have had artistic experience or are sufficiently familiar with handling pencil and pen, you may elect to spend less time with these exercises. In this event, it is suggested that you still read over the exercise, particularly the "Tips" section. It has been our experience, however, that unless you have been making pen-and-ink illustrations almost daily, some practice strokes, drawing rhythm, and feel are necessary to get back in the groove. Some of the drawings in Exercises 1 and 2 are excellent warm-ups for regaining the feel of drawing with pencil or pen.

Carefully read over the entire exercise before starting any drawing so you will know what is expected of you from each exercise. The tips at the end of each chapter may help you further evaluate and improve your drawings. Figure legends within the chapters also contain useful information. Read everything in your search for clues to artistic success.

EXERCISE 1

Purpose: To develop facility with a pencil; to determine and develop an individual "stroke" for free-hand straight lines and arcs.

You will need to use a hard lead pencil for this exercise. The marking on your pencil should read H to 4H (see Annotated List of Supplies, Pencils). This type of hard lead does not wear down as quickly as softer leads.

Drawing 1

Draw a series of 6 to 10 straight parallel lines, 1 to 2 inches long. Place the lines close together, one under the other. Number each set of lines, beginning with 1. Complete at least 75 sets of lines. (See Figure 1.1.)

You are striving, first of all, for an easy and comfortable pencil stroke. Try also to achieve the following goals: straight (nonwavy) lines; parallel lines; uniform thickness of lines; lack of hooks, curlicues, or differences in starting and stopping points.

In order to help develop the best stroke for you, vary the position of your arm and hand, and the manner in which you hold your pencil. (See Figure 1.2.) By the end of the exercise you should have determined how best to position arm, hand, pencil, and paper to achieve the stroke that will give you the smoothest, most uniform,

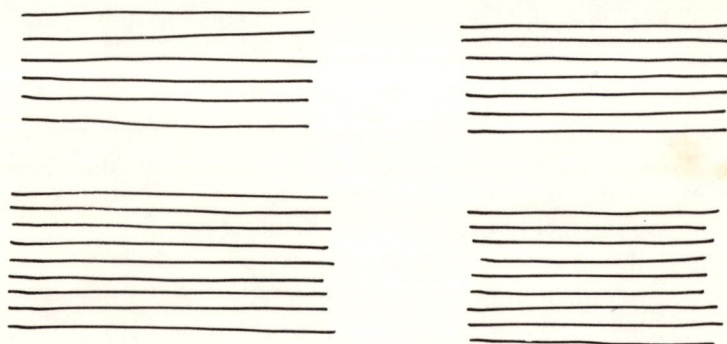

Fig. 1.1. Sets of parallel lines. Notice within and between sets the differences in width and convergence of lines, waviness, and hooks, curls, and unevenness at line ends.

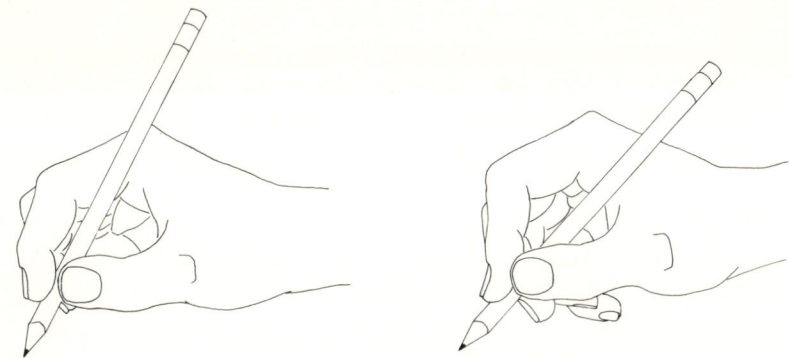

Fig. 1.2. Slight differences in the position of a pencil in the drawing hand may produce limitations on the type of stroke. While there may be two or three relatively comfortable holding positions, try to find the position that produces your best drawing stroke.

most solid, and straightest line. Uniformity might be checked by going over the same line twenty-five times. If you can master repeat lines so that in twenty-five tries the width of the straight line does not vary (over one and one-fourth the width of the original line), you will have found your stroke.

Once you have found your stroke, you should now consider the product of your stroke, the quality of the lines you have drawn. Compare the numbered set of lines and answer the following questions (self-critique of Drawing 1):

1. Are your lines parallel? Is there any improvement from your first set (number 1) to your final set (last numbered set)?
2. Do the lines have uniform thickness (line width)? Can you identify regular variations in the thickness between the first part of individual lines and the last part (an indication of different pressures applied to the pencil or of twists in your stroke)?
3. Are your lines the same color intensity? Compare various line sets (numbers 1, 25, 50, and 75) to see if you can detect variations. Sometimes crossing your eyes or otherwise deliberately observing the line sets with vision slightly out of focus will emphasize differences in the color intensity of the lines.

4. Do you feel you have mastered a stroke for drawing short, straight line segments without using a straightedge (rule) to help you? While you will always have rules available for drawing tasks, biological subjects seldom have such straight lines that the use of a straightedge is undetectable.

Drawing 2
Draw a series of curved parallel lines (arcs) approximately 2 inches long. Again using a hard lead pencil, complete seventy-five combinations, numbering each series. (See Figures I.3, I.4.)

Strive for a comfortable stroke with such a controlled repeatability that you can duplicate the same arc time after time. Test your arc stroke by rotating the paper after a single arc and joining the second stroke to the first. Ultimately, you should be able to construct a nearly perfect circle composed of a series of short strokes or arcs.

Self-critique of Drawing 2:

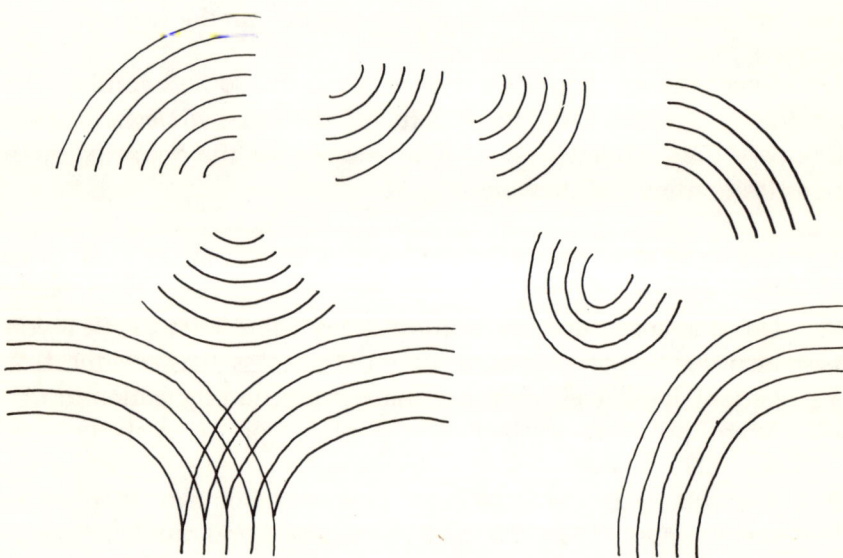

Fig. 1.3. *An example of a series of curved parallel lines drawn while determining a comfortable pencil stroke.*

Fig. 1.4. An example of additional parallel arcs and line segments drawn to practice pencil strokes.

1.   Ask yourself the same questions as for Drawing 1.
2.   Are there any hooks or slight irregularities of angles or a dotted appearance at the beginning or ending of each line? Such inconsistencies in stroke pressure are generally due to hurrying and lack of caution; you should concentrate a little harder on developing consistency.

Drawing 3
   You are now ready to draw free-hand circles and other geometric shapes with a pencil. (See Figure I.5.) Use both the straight and the arc stroke.

Pay close attention to where lines join; make certain both lines have the same width and color intensity, and that there are no re-curved hooks or noticeable overlaps. In other words, strive to disguise where you have joined two lines together (in either straight or arc strokes) so that the line junction cannot be detected. Attempt parallel circles inside and outside your original circle (concentric circles). Strive for uniform line thickness, perfect junctions of lines, and intensity so uniform that you cannot detect any variations in color where the circle was started. Number each attempt for ease in discussing during critique sessions.

Self-critique of Drawing 3.

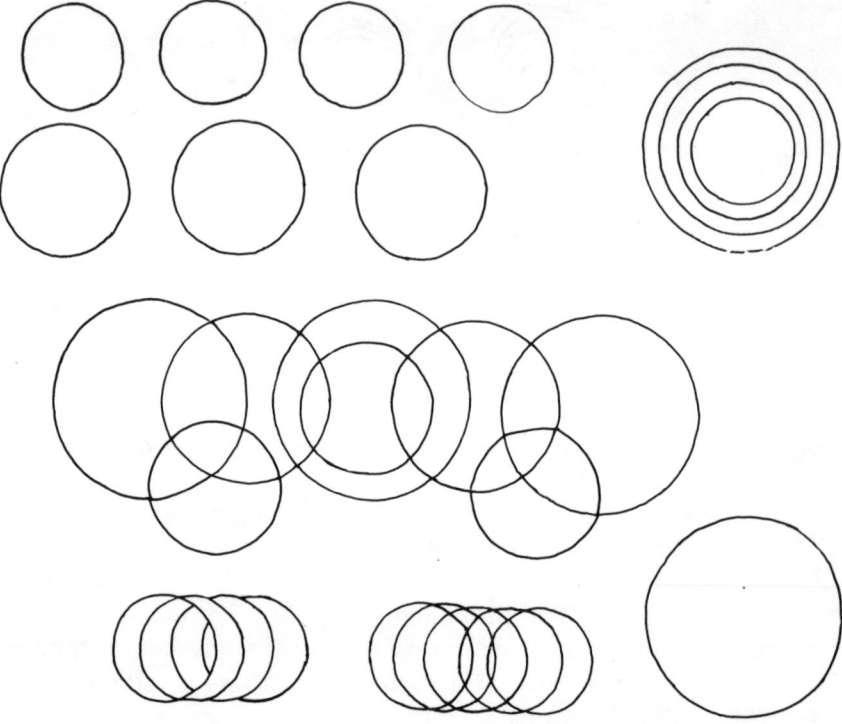

Fig. 1.5. Free-hand circles and geometric shapes using curved and circular strokes.

1.   On circle figures, can you detect where lines join?
2.   Have you achieved a nearly perfect round circle?
3.   Are junctures between straight and curved lines undetectable?
4.   Are lines of concentric circles parallel?

Drawing 4

You are now ready to demonstrate your skills by drawing a series of wavy parallel lines with both curved and straight alternating segments. (See Figure I.6.)

This should put to the test your best straight and curved strokes and line junctions. Label your first and last attempts so they can be compared.

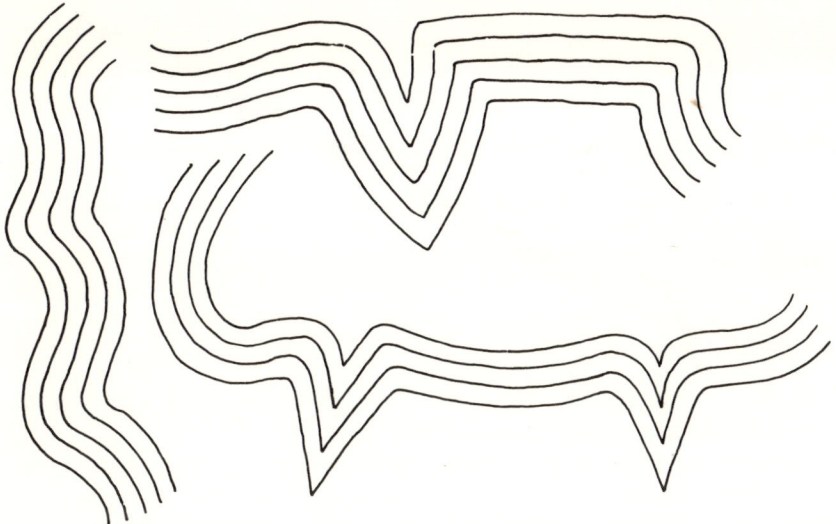

*Fig. 1.6.   Drawing parallel lines alternating straight and curved
segments gives practice in combining strokes.*

Self-critique of Drawing 4.

1.   Do you note differences between the first and last series of
     wavy parallel lines?
2.   What difficulties did you experience in making wavy parallel
     lines? How might you overcome these problems?

EXERCISE 2

Purpose: **To develop facility with a pen; to develop a stroke;
to become skillful at making pen tracings of pencil lines; to overcome
faults at line junctions.**

Stroke exercises similar to those with a pencil will be under-
taken using a crow quill pen. (See Figure I.7.) Familiarize yourself
with this pen, used to draw very delicate lines.

New pen points are coated with a protective fine film of oil,
so that the ink does not flow evenly over their surface and between
the nibs. Several methods can be used to prepare new points: (1)

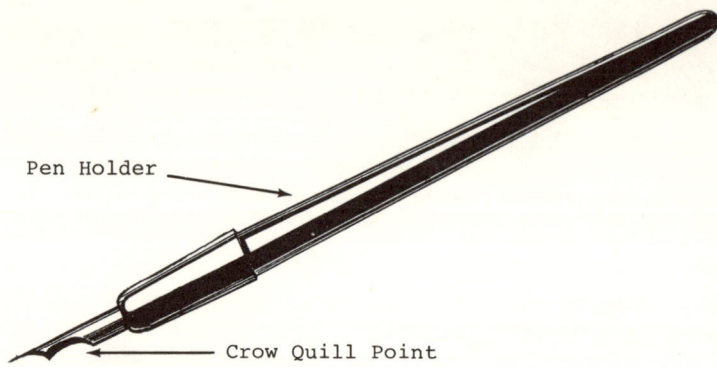

Pen Holder

Crow Quill Point

Fig. 1.7. Anatomy of a crow quill pen. (See Chapter 4, Tip 1, which discusses highlights and critiques this drawing.)

soaking in saliva, (2) holding briefly in a match flame, or (3) repeatedly dipping and wiping in ink prior to use. Make several lines on scratch paper with a new point before use on a half-finished drawing to avoid splatters and blotches. You will be more successful using

Fig. 1.8. Applying ink to the lower surface of the pen point. (See suggested critique of this drawing in Chapter 4, Tip 1.)

pen points if you wipe them after every second or third application of ink with a lint-free cloth or chamois skin (see Annotated List of Supplies) and apply the ink to the bottom surface rather than dipping the entire point in the ink supply. Also, ink flowing off both upper and under surfaces of the pen point makes a thicker, less easily applied line. Most India ink bottles have applicators attached to the stopper that aid in placing a single drop of ink on the underside of the point. (See Figure I.8.) Too much ink may blob up and roll off the point, causing blotching. Too little ink may not flow off the nibs, or may require additional applications.

The two nibs on the tip of the crow quill point are easily bent or misaligned in rough cleaning. Crow quill points are inexpensive; It is wiser to replace a worn or bent point than ruin a drawing.

Drawing 5

Draw a series of straight parallel lines using a slow, deliberate stroke. (See Figure I.9.) Then draw several groups of evenly spaced parallel lines using fairly rapid, short strokes.

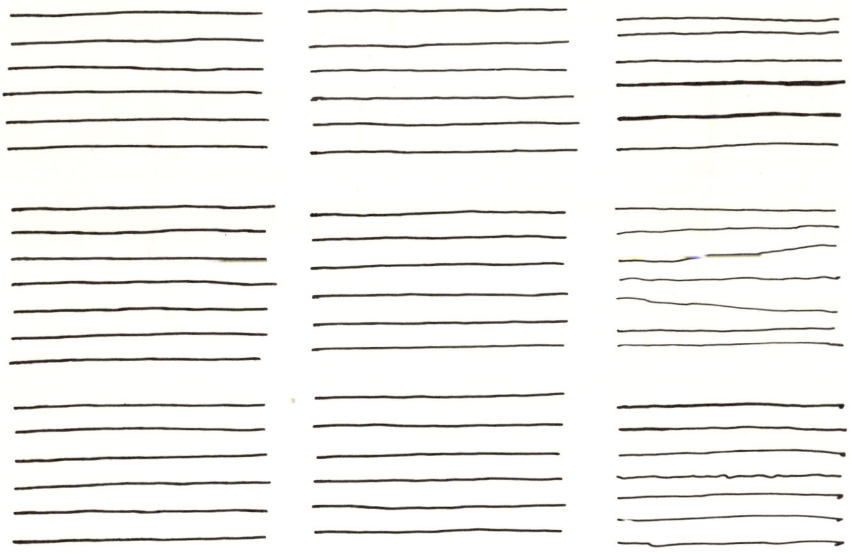

Fig. 1.9. Straight parallel ink lines drawn while determining a comfortable pen stroke. In the third column of figures are found six common drawing faults. Can you spot them all?

Start the strokes boldly and end them with diminishing pen pressure, which should taper the line size. Strive for straightness and evenness in line, but remember that you are working for pen stroke familiarity and comfort.

Try stroking with flexion (bending and tightening) of the hand and forearm as well as with extension. You may find more comfort and ease in one of these types of motion. You may also notice differences in pen pressure during your stroke that will cause the nibs to spread (or contract), varying the width of the line. Holding the pen too loosely or too tightly may cause unequal pressure on one or the other of the pen nibs, resulting in a ragged or feathered line, particularly on one edge. (See Figure I.I0.)

Self-critique of Drawing 5:

1. Did your stroke visibly improve from your first to last set of lines?
2. Are your lines of standard and equal thickness?
3. Can you detect gaps or ragged, sketchy, or feathered lines?

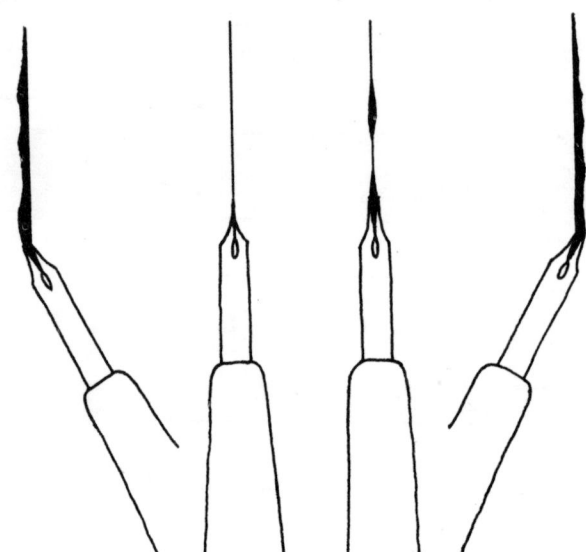

Fig. 1.10. Pen point and paper junction showing separation of nibs with improper pressure and/or angle and the line faults that result.

4.   Are your lines parallel, with a neat, crisp appearance?
5.   Do you feel you have mastered a stroke for drawing short,
      straight line segments without using a straightedge to assist you?

      The following drawings develop skill in ink tracing over pencil
lines, a procedure that the scientific illustrator uses constantly.

Drawing 6

      Using an H to 4 H pencil, draw a short line series and a
long line series of straight lines, arcs, and gracefully sweeping
curves. (See Figure 1.3.) You should draw ten to twenty sets of
each. After completing a practice page, trace the pencil lines
with India ink. Then draw a number of shapes in pencil. (See
Figure 1.11.) Trace in ink, striving for line consistency.

Fig. 1.11. Example of shapes drawn first in pencil and then traced
in ink to practice exactness in tracing. Note that several different
strokes are required.

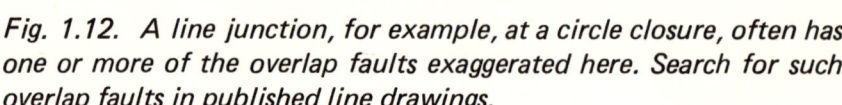

Fig. 1.12. A line junction, for example, at a circle closure, often has one or more of the overlap faults exaggerated here. Search for such overlap faults in published line drawings.

Pay particular attention to the exactness of tracing so that little or no pencil shows beneath the ink lines you have applied. Also, with longer lines, the junctions of your pen lines (the interfaces between pen strokes) become important. This is the area where many faults appear in line illustration. (See Figure 1.12.)

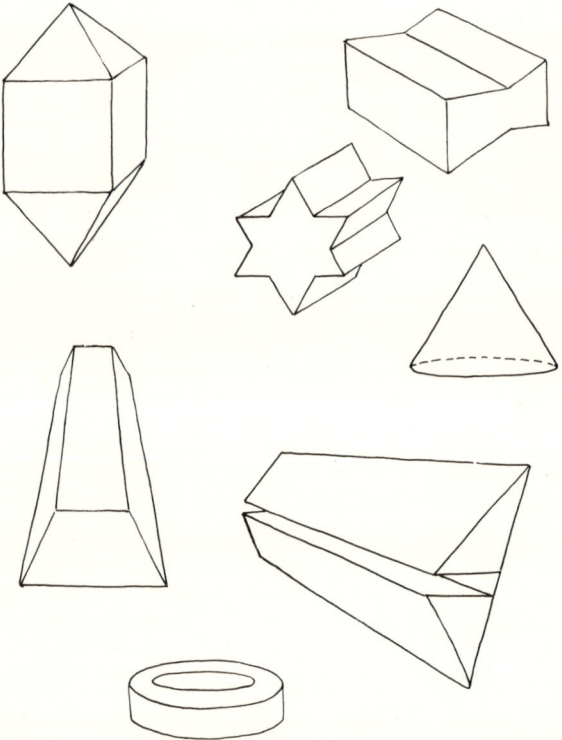

Fig. 1.13.   Inked pencil line tracings of solid geometric figures.

Self-critique of Drawing 6:

1. Are your ink lines of equal thickness and without hooks?
2. Are any pencil lines visible beneath the ink lines?
3. Can you detect juncture problems in the longer lines where more than one ink stroke was used?

Drawing 7

  Draw a series of three-dimensional (solid) geometric figures: boxes, pyramids, and circles. (See Figure 1.13.) Draw them in pencil first, then trace the pencil lines in ink. Be certain to concentrate on line overlaps and junctions.

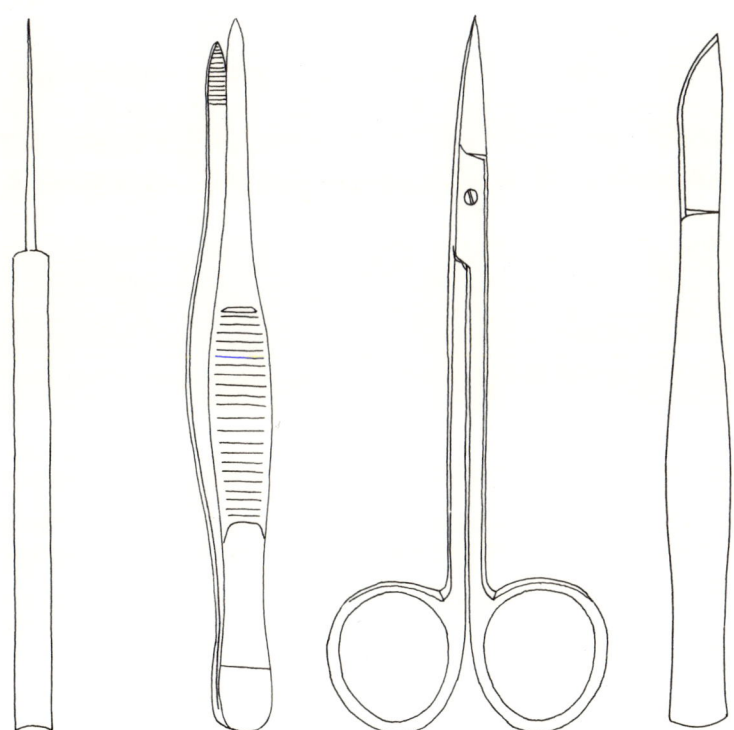

Fig. 1.14. Ink outline drawings of some common laboratory instruments. Can you see line junction faults in these figures?

Self-critique of Drawing 7.

1. Were you able to trace over the ink lines without jumping off line?
2. Observe the line junctions. Are they clear and defined, or are they ragged with overlaps?
3. Are corners joined squarely, or can you detect gaps, or over-lapping lines?

Try some outline drawings of common objects. (See Figure 1.14.) Continual practice at free-hand illustrations helps develop your techniques of line drawing and accurate proportions, building confidence.

The following section progresses to slightly more complex line designs and patterns to increase your facility with the pen.

Drawing 8

Construct a series of about twenty postage stamp-sized penciled squares in which to practice ink strokes and designs of several kinds. vertical parallel lines, slanting parallel lines, horizontal parallel lines, wavy parallel lines, cross-hatching, hooked strokes, overlapping scales (like shingles on a roof), golf ball dimples, moon craters, and contour lines. (See Figure 1.15.) In short, practice with your pen as many patterns as possible in each square, noticing that fairly simple lines and designs can result in pleasing effects of texture, contour, and tone.

As you practice new strokes, try varying the pressure on the pen point, which will open or close the nibs and vary line thickness. One stroke of this type easily achieved by a beginning illustrator is the "short-hair" or "bristle" stroke. (See Figure 1.15C.) Upon contact with the paper, spread the nibs; a quick short lifting motion makes a short tapered line resembling a bristle, thicker at the base than the top. With practice, a fur coat of very realistic hairs can be drawn.

Wavy, contour parallel lines may also be varied in thickness to achieve different effects. (See Figure 1.15A,B,D.)

Fig. 1.15. Ink stroke practice patterns and designs. Many biological textures, contours, and tones may be achieved with such relatively simple patterns.

Self-critique of Drawing 8:

1. Are your ink lines sharp and clear? Are any patterns in your squares too sketchy?

2.   Do you note any faults such as overlap lines, hooks, or uneven lines?

3.   Have you achieved some new and interesting designs or unusual textures or tones?

Self-critique of Exercise 2:

Now that you have completed Exercise 2, do you feel more confident with pen and ink? What problems are you still having? Discuss these with your instructor. If you have continued problems, repeat all or parts of Exercise 2. In fact, Exercise 2 is a good one to warm up with before attempting more complex illustrating. It will help you to recover your stroke if you have been away from illustrating for any length of time.

TIPS

1.   Cleanliness
       The most precious ally of the illustrator is cleanliness. Extreme care with paper, drawing surfaces, and instruments will avoid some problems with the final drawing product or reproductions of it. During pencil drawing, protect the paper from accidental smudging. It is sometimes desirable to keep a small piece of paper under the drawing hand where it contacts the paper. Not only does this help prevent line smear, but it keeps sweat and grease from the illustrating surface. The latter may seriously affect the absorptive qualities of the paper or its ability to hold graphite.

2.   Erasing
       Rubber or Artgum erasers are suitable for correcting most pencil lines and shadows. A paper or metal shield is sometimes used to protect nearby areas of the drawing when corrections are made. Some types of illustration board may have their surface textures altered slightly by erasing, with the result that repenciled lines or shadows may be darker, so adjustments may have to be made. Unsightly smudges or smears in areas of highlights or white areas surrounding the drawing may have to be opaqued out (see Annotated List of Supplies) if the drawing is to be reproduced.

3. Protection of drawings

Penciled drawings may be protected by spraying the final drawing with thin, clear fixatives available at most art supply dealers. These may be acrylic plastic finishes or nonglossy (matte) finishes that permit changes or corrections to be made on the drawing. (See Annotated List of Supplies.) Even hair sprays can be used (avoid those with lanolin or colored tints), lightly applied to the surface to avoid running. The nozzle of the can should be held 10 to 12 inches away from the surface, so that the mist gradually covers the drawing. Before using on a drawing, be sure to try out such a spray on a small sample of the illustration board used.

4. Pen point care

Always shake off excess ink, or lightly touch the blob of excess ink on the point to the inner lip of the ink jar so that it runs off the point back into the bottle. Attempting to use a point with too much ink is flirting with the "great blob," a disaster familiar to every illustrator. As suggested in the text, wipe off pen tip with a lint-free cloth or chamois skin after every second or third ink drop used. Cleaning pen points periodically in diluted (50%) household ammonia lengthens their usefulness and ensures a smooth flow of ink off the point.

5. Ink

Drawing inks should always be absolutely black and waterproof. Good quality inks are more expensive, but they repay the user by uniform quality, adhesion to surface on erasure and cleaning, and other desirable properties. (See Annotated List of Supplies.)

6. Ink corrections

Corrections in ink lines or dots can be made by chipping the ink away with a scalpel or razor if it is a small area, or by covering it with an opaque white if it is a larger area. Opaque retouching and correcting white pastes or fluids are available at all art supply stores. These should be allowed to dry thoroughly (heating with a lighted bulb hastens the process) before reapplying ink to the surface. The opaque white is usually applied in a thin layer with a fine brush.

2

Views on Perspective_____

Although "seeing is believing" is usually true in most fields, every biological illustrator is aware that illusions are necessary in drawings. Most of these are not intended to delude the viewer, but are instead based on the fact that we are usually attempting to draw three dimensions on a paper with only two dimensions. In order to do this, some techniques of perspective and foreshortening must be used. In this way, the roundness of snail shells, for example, can be brought out as the shell "disappears" over the horizon of view. How does one draw in pen or pencil the three-dimensional depth of a snail shell? Do all angles of view of the shell require the third dimension?

Exercise 1 deals with elements of perspective and foreshortening and those techniques necessary to help establish depth in a drawing. Read the exercise, including the Tips section, before undertaking the drawings.

EXERCISE 1

Purpose: To introduce and/or reinforce the concepts of using perspective and foreshortening in line drawings to give the illusion of depth.

Perspective deals with the apparent diminution in size of objects as they recede into the distance. Hold a ruler level in front of your eye, pointed away from you. Tilt the ruler upward at its far end. Notice that the far end looks smaller than the near end, and that the sides of the ruler do not appear parallel. This problem of perspective is dealt with in the first part of this chapter.

Almost everyone is familiar with the concept of the "vanishing point" in illustrations. This concept is used when drawing an object that extends to, or a great distance toward, a distant "horizon." An example is a road that stretches to the horizon on a featureless landscape. A drawing of such a road can be accomplished with three lines. (See Figure 2.I.) Does the diagram on the left look like a road heading to the horizon over a hill?

Drawing 9

Draw a horizontal line, in pencil, across a page, half the distance from the top. This line will represent the horizon. From a single point on this line, called the vanishing point, draw two lines toward the bottom of the page, each about 4 inches long. (See Figure 2.I, right.) Draw the lines so that they end about 1 1/2 to 2 inches away from each other. With a little imagination you can visualize this as a road stretching from below the level of the observer to the distant horizon. You will notice that the two lines representing the sides of the road look parallel, though, in fact, they are not. The reason for the parallel appearance is perspective, achieved by the use of a vanishing point. The vanishing point is where the parallel lines converge (touch).

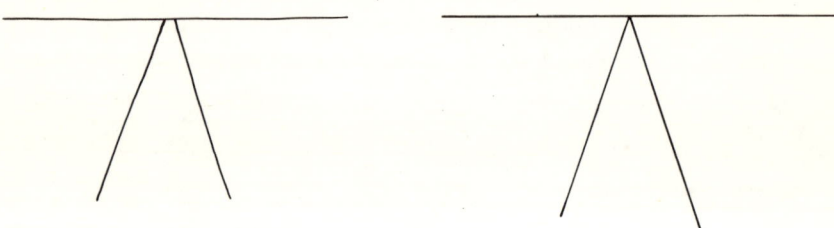

Fig. 2.1. Three-line roads and horizons.

Drawing 10

Make another road on your landscape, but this time place the vanishing point dot about 1/2 inch above the horizon line. (See Figure 2.1, left.) The two lines representing the sides of the road should end at the horizon, so that they would have converged at the artificial vanishing point in the sky. Notice that the horizon appears much closer to the observer in this drawing. This is because the road, with which we all have some familiarity, disappears over the horizon in the second drawing, before the road edges converge (about 1/4 inch of space between them).

Drawing 11

Make another pencil drawing in which the road is converted into a railroad by making the single line representing the road margin into rails. This can be done by adding a second line to each side and by adding crosspiece railroad ties. If you have done this properly, both the spacing of the ties and their detail, as well as the rail width, will diminish or disappear with distance. This line drawing again illustrates the use of a single vanishing point. (See Figure 2.2.)

As a second part of this line sketch, add some fence posts or telephone poles adjacent to the railroad to further emphasize the single vanishing point concept. If telephone poles are used, their tops may well be above the eye level of the observer; so the parallel line joining their tops will extend above the horizon line. The fence posts, on the other hand, because they are shorter, will no doubt still be below the eye level of the observer, and a line joining their tops to the vanishing point will end considerably below the horizon line. Make certain that the fence posts and telephone poles are at a right angle to the railroad ties, or they will appear to lean in an unrealistic way.

When you have completed your line diagram, complete the diagram with a free-hand ink tracing.

Cut out this drawing and fix it with rubber cement to a heavier, more supportive Bristol board or poster board (see Annotated List of Supplies).

Drawing 12

Using a 4H pencil, draw a cube using the vanishing point. This may be done by first drawing a 2-inch square, then placing

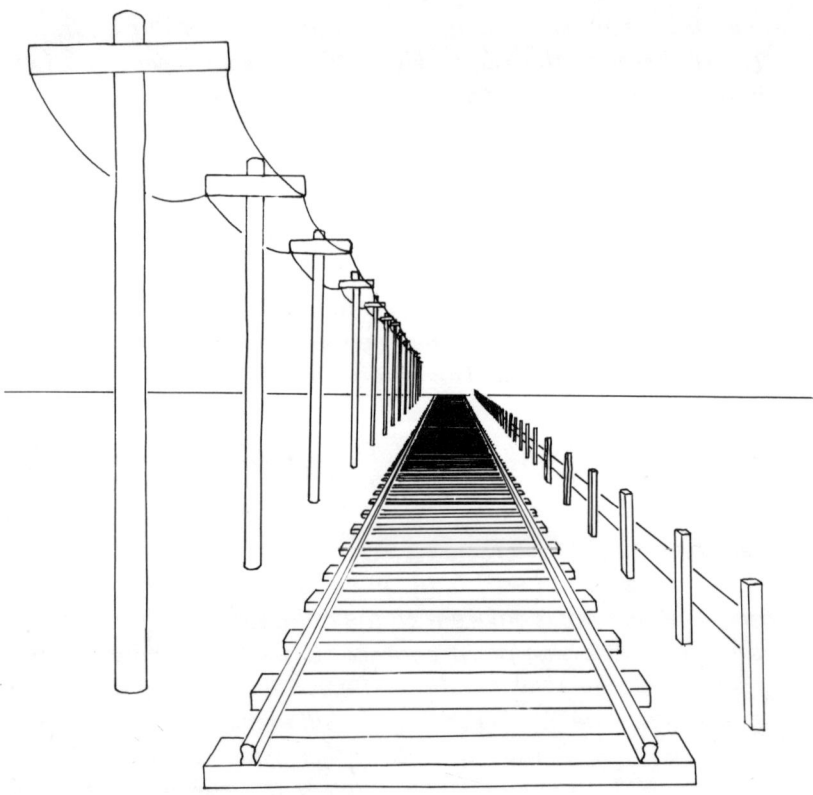

Fig. 2.2. A drawing of railroad tracks, fence posts, and telephone poles using a vanishing point. In this case the vanishing point is above the distant horizon. Had the vanishing point been on the horizon, the lines following the tracks, fence posts, and telephone poles would have joined there. Comment on the darkened railroad ties in the distance (should the linear detail have been eliminated?) and the effectiveness of the diminution of the telephone poles and lines and fence posts with distance.

a dot representing the vanishing point 1 inch above and 2 inches to the right of the square. (See Figure 2.3A.) Next draw three lines from the vanishing point to three corners of your square. (See Figure 2.3B.) Finally, across these lines draw two lines that are parallel to the square, and partway to the vanishing point.

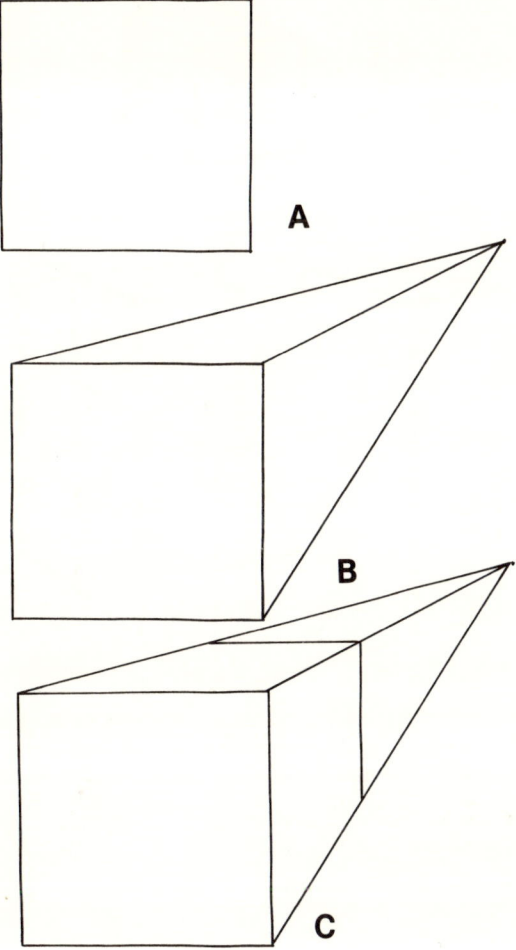

Fig. 2.3. The sequence of drawing a cube using the vanishing point concept.

These cross lines contribute to the depth perspective that gives the illusion of a solid figure. (See Figure 2.3C.)

Your completed cube may not seem accurate, however, if you neglect to foreshorten the cross lines. Foreshortening is a technique of proportionately shortening to give an illusion of depth. A cube

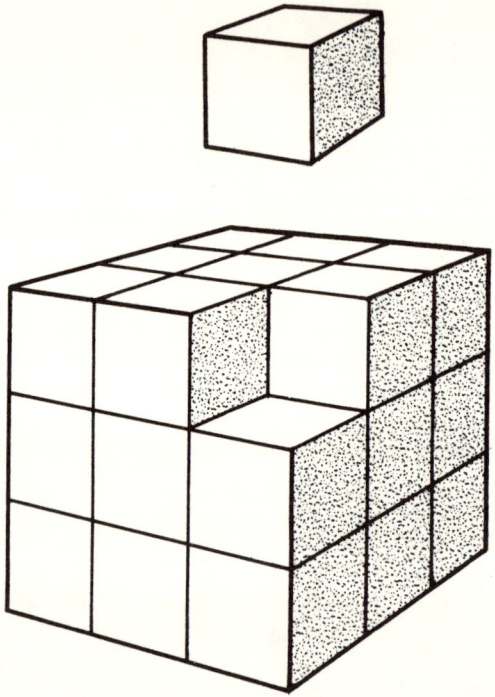

Fig. 2.4. Perspective in cubes drawn using a (lightly drawn) vanishing point. Lines were inked with a ruling pen, the vanishing point lines were erased, and a patterned acetate sheet overlay was used on the shadowed surfaces. (See Annotated List of Supplies, Acetate sheet overlay.)

that has, by definition, the same height, width, and breadth would need to be foreshortened in order for its depth in a drawing to appear equal to its height and width. (See Figure 2.4.)

Drawing 13

Try drawing a cube with the back top and side lines the same length as the front top, side, and bottom lines. Does this box look like a true cube? What is the actual shape of the resulting figure? Trace all your pencil cube drawings with ink.

In Figure 2.3C, the cube appears to be to the left of the viewer. This is due to the placement of the vanishing point to the right of the cube. If the vanishing point was placed to the left of the cube, the cube would appear to be to the right of the viewer. Test the effects of varying positions of cube and vanishing point placement by drawing ten 1-inch squares. For each square, place a single vanishing point somewhere outside its dimensions; include some placed above the square and some below. Note how the cubes appear in respect to the viewer. Finish the pencil drawings of the final cubes in ink.

Locate the vanishing points on the shapes in Figure 2.5 by extending the lines with a ruler. Note that the upper shapes have two

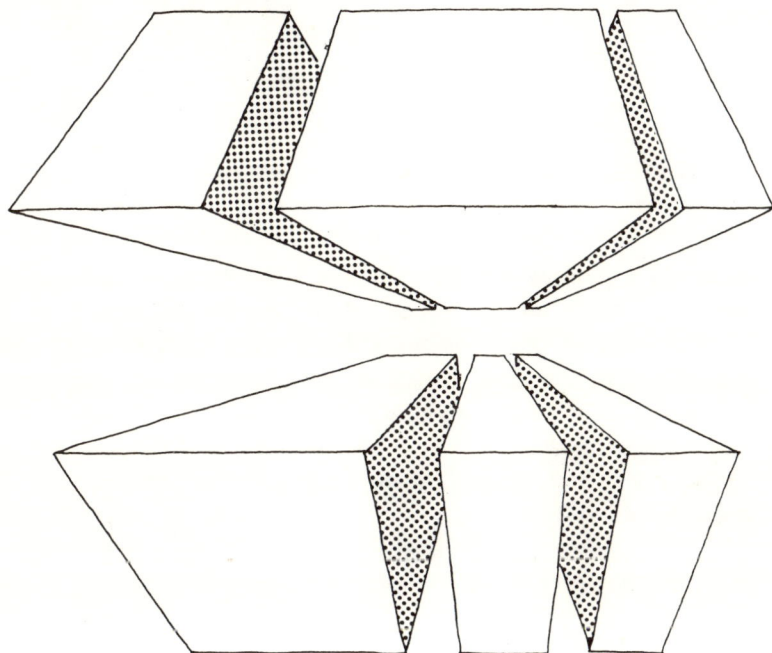

Fig. 2.5. Three-dimensional shapes drawn with the aid of multiple vanishing points.

vanishing points (one above and one below). The lower shapes also have two vanishing points, one of which is shared with the upper shapes.

Drawing 14

In pencil, draw several dime-sized circles. By adding vanishing points and lines, convert them into "pipes" of varying lengths. Use a common vanishing point for several circles and

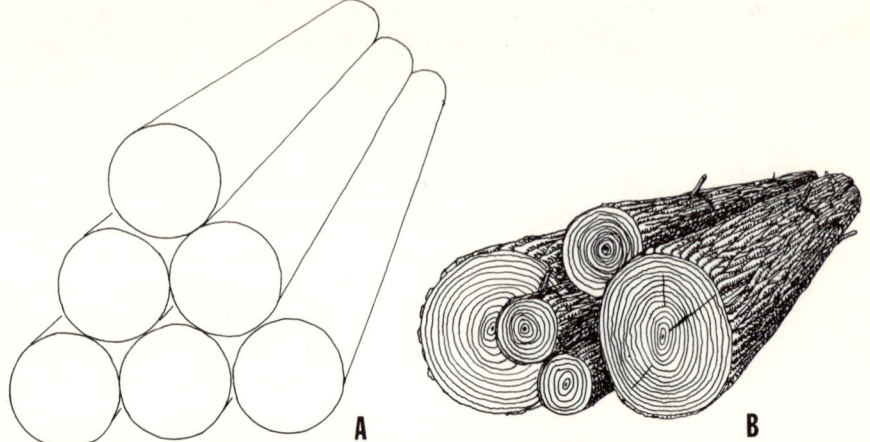

Fig. 2.6. Stacked pipes and logs drawn in perspective.

extend your pipes toward this. The effect you end up with is a series of pipes in space, or, with proper placement of the circles, a stack of pipes. (See Figure 2.6A.) With only slight modification, the same principles can be used in drawing a series of stacked logs. (See Figure 2.6B.)

Complete the stacked pipes or tubes in ink. The line you use to complete the end of the pipe must be a curved line. Do you know why?

To develop additional skill, make all drawings free-hand, even if drafting tools (compass, straightedge, ruler, ruling pen) are available and would probably be used for finished work of this sort.

A true perspective drawing involving curved outlines and surfaces is more complex than Drawing 14. For example, see Figure 2.7 of coins drawn in perspective. The round coins change to ellipses as they turn around a vertical or a horizontal axis.

The execution of a perspective drawing involving curved figures and surfaces is always more difficult than those of shapes with straightedges. Drawing hemispheres is a good example of this problem.

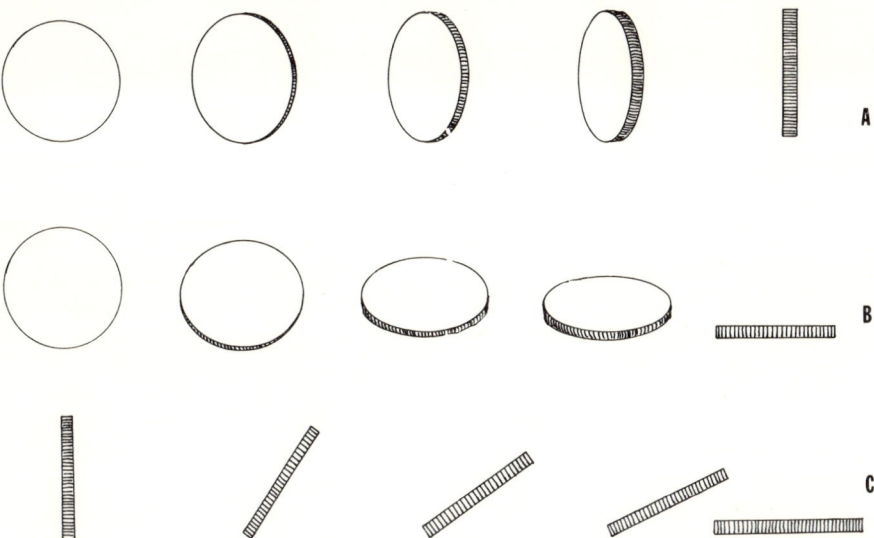

Fig. 2.7. Coins in perspective. (A) Twisting on a vertical axis. (B) Revolving on a horizontal (helical) axis, or simply dropping from vertical to horizontal position. (C) Edge-on view, vertical to horizontal axis.

Drawing 15

Draw one-half an orange or grapefruit by using the following steps:

In pencil, lightly draw a circle. Bisect the circle with a straight line. Draw an ellipse around this centerline. (See Figure 2.8A.)

Carefully erase the right side of your drawing so only the left hemisphere remains. (See Figure 2.8B.) By erasing the centerline and inserting a center point with radiating lines and a "rind" or "peel," the effect achieved is that of half a citrus fruit. (See Figure 2.8C.)

Make several additional drawings of hemispheres by varying the width of the ellipse drawn around the line (diameter) bisecting the circle. Notice that the width of the ellipse determines the tilt of the hemisphere. Notice also that the ellipse within a circle (Figure 2.8A) can easily be converted to a solid orange from which a section has been cut. To avoid optical

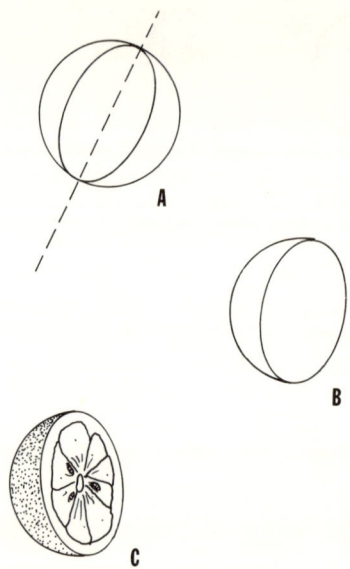

Fig. 2.8. Steps in drawing an orange, using a hemisphere.

illusion it would be best to add details to the cut section of the figure. Try this now by making several drawings illustrating oranges with quarter sections removed.

Self-critique of Drawing 15:

1. Did you have any difficulty with line junctions?
2. Do your hemispheres resemble citrus fruit?
3. Do the lengths of the ellipses and/or the widths of the ellipses have any effect on the size of the total hemisphere?
4. Examine Figure 2.8C.
    a. Is the execution of perspective well done (does the figure look three-dimensional)?
    b. Is the execution of the subject well done?
    c. Note the rind or peel; compare texture with that of the pulp. Is the pulp realistic? (Remember, accuracy is the important quality of biological illustration.)

Foreshortening and perspective are related, but they are not the same thing. As we have noted, perspective is involved with vanishing points and the decrease in size of objects as they recede into the distance. In foreshortening, we portray an illusion of depth by a series of techniques including overlapping, varied line widths, line breaks, and/or distortion of proportions. One or all of these techniques may be appropriate for giving depth perspective to the drawing. (See Figure 2.9.)

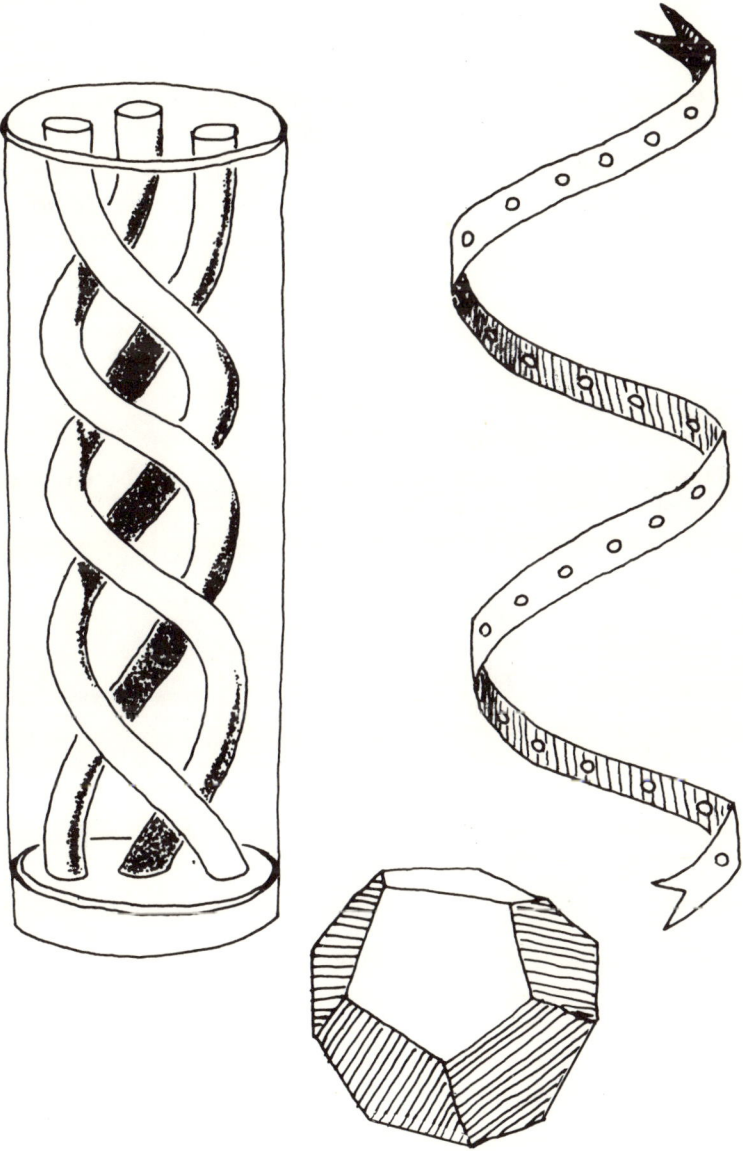

Fig. 2.9. Drawing three-dimensional figures helps develop techniques of perspective and foreshortening.

For the following drawing, which demonstrates foreshortening, you will need an L-shaped object such as a piece of glass tubing bent in a right angle.

Drawing 16

Draw a side view of an L-shaped object, such as bent glass tubing. Then make a series of drawings, all with the same dimensions, of the object placed in several positions. (See Figure 2.10.) In doing this you will need to foreshorten one arm of the object as it turns closer to your angle of view. Illustrating the elbow of the "L" in these varying positions, particularly without the aid of shadowing, is one of the most difficult problems of foreshortening.

Self-critique of Drawing 16:

1. Are your lines the same thickness?
2. Are your lines parallel?
3. Have you used a circle or an ellipse to represent the opening of the glass tubing?

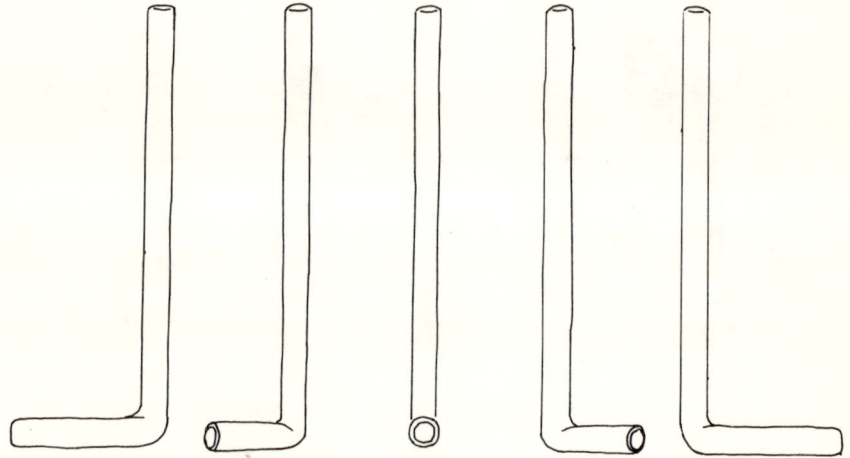

Fig. 2.10. Foreshortening in drawing right-angle glass tubing.

4.   Is the concept of foreshortening well illustrated?

Foreshortening can even be achieved in linear drawings. Notice in Figure 2.11 the ways in which this is achieved: varying line widths (the wider the line, the closer it appears to the observer), and line breaks (a line will appear to go behind another if it is broken on both sides of the line in the front position), as well as vanishing point.

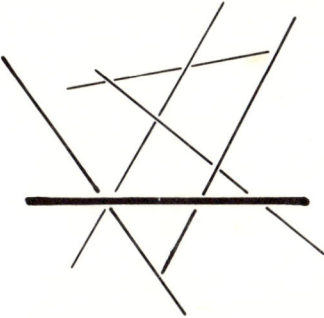

Fig. 2.11.   Entwined "toothpicks" showing line breaks and depth perspective. Which stick seems closer to the observer, 1 or 2? Why?

Drawing 17
        Draw several "toothpick" lines lying one on top of another and intertwined. (See Figure 2.11.) Notice that the larger the line breaks (in toothpicks in the background) the farther away they appear to be. Try holding the toothpick drawing at arm's length. Do you notice any depth perception?

Examine Figure 2.9. One of the geometric figures has three entwined wires in which line breaks were used to show depth. Does this device add to the perspective of the drawing?

Drawing 18
        Draw a small series of circles representing a string of beads. (See Figure 2.12.) You may show small parts of the string between adjacent beads. In a second drawing, bend a portion of

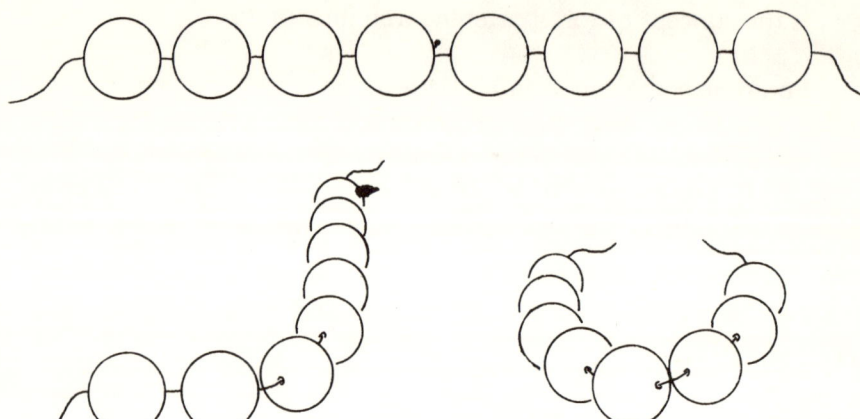

Fig. 2.12. Drawing of a string of beads showing perspective in spheres.

the string, so that the beads appear to line up going away from the observer and slightly uphill. Note that in doing this, you will have to foreshorten the string in order to gain some perspective.

Now that you have made numerous examples, in Drawings 9 through 18, of techniques to illustrate depth, select several of your best efforts. Cut these out and arrange them on one or two Bristol board plates as examples. They may be glued with rubber cement or dry mounted to the heavier mounting board. (See Chapter 2, Tip 4.)

Compare the length of the right and left limbs in Figure 2.13. Foreshortening was achieved in the right arm by a combination of features: overlapping and proportional shortening.

TIPS

1.   The angle of view of the subject

In drawing an object, the illustrator may need to provide specific views. For example, custom may dictate in some biological areas that full dorsal (upper) or ventral (lower) views be given in line illustrations. If the choice of view is unrestricted, or if there is a desirable feature to emphasize, the illustrator may select the most appropriate view. This is often

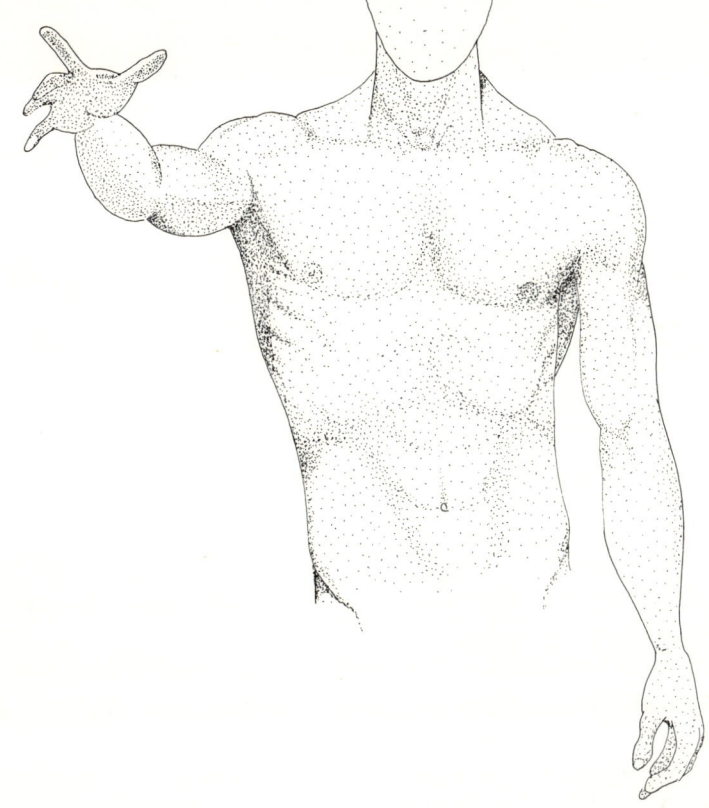

Fig. 2.13. An outline figure of human torso, right hand and forearm extended and foreshortened.

done by eliminating those views or perspectives that are more difficult to illustrate. For example, when illustrating a fingernail on a human finger it would be best not to show it with the finger pointed directly at the viewer. The difficulty with this view may be even more dramatically illustrated by having a friend point a finger at your eye with the tip held about 1 inch away from your eyeball. This position, with its resulting problems of perspective and foreshortening, is much more difficult to draw than when the finger is several inches, or 3 feet, away from the observer. Examine Figure 2.14. When the can is held close to the eye (Figure 2.14A), its upper and lower surfaces appear strongly curved outward. The farther away from the eye

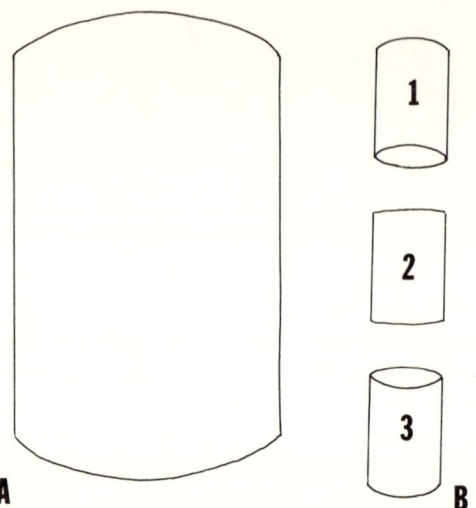

A   B

Fig. 2.14. Demonstration of parallax in line drawing of cans. (A) Upper and lower curved surfaces indicate the can is "close" to the eye of the observer. (B) The further away the object, the less distortion caused by parallax.

(Figure 2.14B), the smaller is the curvature, the less evident the distortion; this is a much more easily illustrated view.

To keep the specimen in the same view, it is often helpful to mount it on a small lump of modeling clay or soft wax, or to place it against cut eraser props. (Having a gooseneck lamp or other set light source coming from the upper left may also help establish proper highlights and shadows, useful in establishing perspective and foreshortening.) A small wire or pencil sight may be used as a point of reference to help establish a head position, enabling the illustrator to view and measure the object from a set position.

2.   Size of drawings and reducing drawings

After selecting the best angle of view of the subject and one that minimizes distortion, you have to decide on the size of the drawing. This should be done with the realization that the results will be in two dimensions even though the specimen is viewed in three dimensions. Critical linear measurements must thus be taken from a two-dimensional perspective. Perspective foreshortening will add the third dimension but will not affect these measurements.

If drawn for publication, the product should always be done larger than will be used in duplicating. In other words, the illustrator should ordinarily plan to make the drawing two to three times the desired final dimension. Many illustrators plan their drawings to be not more than twice the final reduced size, but reducing to 2/3 of the original size is very common. Reducing to less than 1/4 the original size makes the original drawing so large as to inconvenience printers.

Reductions are regularly made of line illustrations to decrease line widths and to deemphasize stipples and other techniques to suggest tone and texture (see Chapter 3), which enhances the final effect. That is, if stipple dots (see Chapter 4) are too large, they may look like dots to the viewer and not like the tone or shadow they are meant to suggest. After reducing, however, the total effect of a number of very small dots may be one of shadow (or gray). (See also Chapter 6, Tip 2.)

3.   Drawing proportions

Proportions of figures may be measured by rules, calipers, or dividers. In some cases they may be more accurately obtained by using optical instruments such as a camera lucida attachment on a microscope (see Annotated List of Supplies), projectors of various sorts, or photographs. Photos may be projected and then traced to the desired enlargement for drawing purposes.

A grid system can be helpful in drawing proportions of an object, or in reducing or enlarging a drawing. The object (or drawing) is placed against or on a piece of graph paper. On an enlarged (or reduced) drawing paper grid, the outline of the object can then be duplicated keeping corresponding grid coordinates. A small piece of card, cut to one or more grid widths, can be used in establishing prominent features of the object. Keeping the same two-dimensional view (not moving your head position, so your view remains the same) is one of the difficulties of using the grid system, but such difficulties are more than offset by its usefulness in drawing to dimension, established by the ratio of object-grid size to drawing-grid size, and the relative ease with which parts of the object can be kept in proper proportion.

Proportions may also be established by a glass-tracing technique described in Chapter 3, Tip 5.

4.   Mounting illustrations on protective backing

Illustrations drawn on vellum (tracing paper) or other thin sheets of paper should be mounted on heavy illustration or poster board. Cut the heavy board with a knife or razor to the desired size. Cut the excess paper from around the individual drawing, but leave ample margins on all sides of the figure. If reproducing the drawing, differences in white paper will not show in photographic duplication; shadow lines of the over-lapping paper may be easily opaqued from the negative if draw-ings have adequate margins. Next, position the drawing on the poster board. The position may be faintly marked in pencil, so the same alignment can easily be made during the paste-up process, and so that you know where to apply rubber cement on the poster board.

Brush a good quality rubber cement on the reverse of the illustration and on the thick mounting board to which it will be affixed. Try to coat the cement evenly; this is particularly im-portant if the drawing is on very thin paper. Do not worry if both surfaces appear to dry out rapidly. Reposition the drawing to the mounting board, realign it at the desired position, and press the two surfaces together. Even though seemingly dry, the two coats of rubber cement will fuse; a roller or burnishing tool (see Annotated List of Supplies) may be used to better seat the two surfaces. After this step, clean off excess rubber cement, or any that seeped out from between the two sheets, by brisk but gentle rubbing with clean fingers. A wad of the semidried cement may be used as an eraser and to pick up excess rubber cement.

The advantage of rubber cement is not only the ease with which it may be cleaned from the drawing, but also that it is possible to separate the two pieces later if sufficient care is used. Other glues cause more problems and are not advised, even though they may seem to work. Some photographic methods of mounting (wax sheets, etc.) are satisfactory. In making a plate of several individual drawings, be certain to determine the layout before gluing to keep from crowding the figures. Leave sufficient space for labels or for legends, if they are planned.

3

Making Light of the Dark:
Shadow, Tone, and Values_____

Camera buffs use the terms "underdeveloped" and "over-developed" in speaking of the tones of their prints. "Much too light!" or "Too dark!" are common criticisms of beginning photographers seeking correct exposures. Similarly, the biological illustrator must decide how dark a drawing should be, that is, what tonal value (degree of darkness; light and shadow) should be set. There are many deciding factors: the texture (quality of surface structure as distinct from its color or tone) of the object, its setting and the light intensity involved, the purposes of the drawing, and the amount and nature of existing primary shadows. Most biological illustrations include all of the details of the subject; so that, for example, a black bear would not be "toned" so dark that limbs and details of fur tracts would not be visible. In other words, a silhouette is not the best choice for a biological illustration, though there are situations where black profiles or likenesses cast by a shadow have their place. (Notice the lack of detail on black portions of the bear in Chapter 5, Figure 5.5.)

Thus, decisions have to be made--how light to make a black bear, or how dark to make a polar bear--in order to achieve the best black and white drawing of both the color of the animal and its anatomical details. A polar bear standing in the shadow of an ice-flow cliff might appear black and structureless to an observer. In order to draw the bear to be recognizable, however, you would have to portray it in a different view and light intensity, rather than in a dark environmental setting.

Primary shadows sometimes provide the key to establishing the tone of the entire drawing. If you can determine the details to be shown in the areas of primary shadows, it is possible to work back-

wards from the darkest areas to the lighter ones and thus to determine how dark (or how light) to make the highlights.

Difficulties of interpreting shadow and tone using a pen-and-ink
medium offer a distinct challenge. Such difficulties and challenges
are just part of an illustrator's growth, and they must be faced and
mastered. Exercise 1, which introduces techniques of shadow and
tone, is not without its challenges. Carefully read over the entire
exercise before doing the first drawing.

EXERCISE 1

**Purpose: To develop techniques using shadow, tone, and
texture.**

Shading techniques can add perspective and detail to your
drawings. There are several ways you can shade (shadow) your
drawings. One method is smudging with a stump (see Annotated List
of Supplies). Lightly pencil in shadows by continuous pencil strokes
applied in the same direction, and then obscure lines by smudging
(rubbing) with the stump. (For clarity, do not try to draw shadows
off the figure--those on the surface on which the object rests).

Shadows can also be achieved with hachures (lines). Hachures
might run lengthwise to the folds (see Figure 3.1B) or in the same
direction as the folds (see Figure 3.1C). Shadowing can also be
accomplished by cross-hatching (see Figure 3.1D); variations in line
widths, number of lines per inch, and single or cross hachures may
be used to indicate intensity in tone and/or shadow. When using
line shading in drawing biological objects, the natural contours of the
body must be followed in order to achieve a natural look. The direction of the hachures and variations in width sometimes add to the
character of the subject.

Drawing 19

*Cut three 1-inch-wide strips from a 3 x 5 card. Fold the
strips at irregular intervals, forming accordion folds. Stand these
folded strips on edge. Arrange a strong light source from the
upper left that illuminates some surfaces of the folds while
casting shadows on the other sides. Draw these strips in a pencil
outline and then ink these outlines. (See Figure 3.1A.)*

Fig. 3.1. Accordion folds in a 3 x 5 card with a strong source of illumination from the upper left, shaded by hachures, cross-hatching, and stippling. Which of these drawings would improve with further photographic reduction? You may wish to check this point with a depression slide reducing lens. (See Annotated List of Supplies, Reducing lens.)

The next step is to add the appropriate shadowing. On your inked outlines, draw examples of each of the types of shadowing discussed above.

Shadows and tone can also be accomplished by using regular and irregular stippling.

Drawing 20

Shadow some ink drawings of accordion-pleated paper strips with stippling, first with pencil and then with ink. (See Figures 3.1E, F. Note also Chapter 3, Tip 4.) While stippling techniques are used in greater detail in the next chapter, it is appropriate to consider this procedure along with others that impart shadow and tone to drawings.

Note that the same size drawings shaded with pencil appear different from those shaded with ink stippling. This is because the variations in gray in the pencil drawing more nearly approach real shadows. Ordinarily, we do not see "stipple" shadows. However, upon observing your results with the aid of a depression slide acting as a reducing lens or by squinting (see Chapter 3, Tip 2), you will notice that as the stippling becomes less apparent to the eye, the effect becomes one of natural shading, the original objective in using this procedure. You will have to use more stipples on your drawing to achieve the effect of shadows if the drawing is to be reduced as much as 50 percent.

Drawing 21

Draw and shadow a chicken egg. (See Figure 3.2.) Make a pencil outline of the egg on a heavy, smooth-finish paper (perhaps Strathmore; see Annotated List of Supplies, Paper). Complete the outline in ink using either your crow quill pen or a Rapidograph. (See Annotated List of Supplies, Pens.) Do not attempt to add details of texture to the figure.

Using vellum tracing paper, trace several outlines of the inked egg figure for use in practicing shading techniques. Do not shade where the egg rests on a surface; confine your efforts to the egg itself. Your light source should be from the upper left side. (See Chapter 3, Tip 1.)

Draw the shadows in pencil first, blending the pencil lines to a continuous gray tone by means of a stump. Are there any secondary highlights (reflections from the surface on which the egg rests that may affect the appearance of the shadows on the

Fig. 3.2.   Photograph of egg with a light source positioned at the upper left of the egg. Note the shadow cast on the egg as compared to the lighted front side.

lower right side of the egg)? Including such secondary highlights often enhances the illustration, in addition to softening the contrast between the darkest shadows and the lightest regions.

When you have completed a pencil drawing shaded to your satisfaction, place a piece of tracing paper over it and make a new drawing using only ink stipples as a shadowing technique. Space the stipples evenly, sparsely at first and gradually darkening the primary shadows to the proper degree by adding more and more dots.

Tone (in black and white drawings, the degree of darkness) is a combination of black lines (or stipple marks) and the intervening white spaces, which establishes a tonal value somewhere between white (on one end of the scale) and black (on the other end). Since pen-and-ink drawings are only in black and white, intermediate shades of tone (for texture or shadowing) can be established only by varying the amount of black on the white surface. If a drawing has a large number of black lines, spaced the width of the ink line apart (so that white spaces are equal to the black lines), the drawing will appear medium gray to the eye--that is, about midway between all black and all white.

In the egg drawing, the pencil or stipple technique was used for shadow effects only. Even though the egg was white in color, you probably were not bothered by the interpretation of shadows on a white surface. However, if the shadowing was overdone--stippling to the point of blackness--it would seriously affect the tone of the egg. A black egg has a much different tone from that of a white egg, and a drawing of one would have to portray this.

Tone establishes the total lightness or darkness of an object. When working with color, this is sometimes called its hue or tint. The hardness or softness of an object, its ability to absorb or reflect light, thus help determine its tone. The texture of a surface, whether it is (for example) smooth, grainy, pitted, or fibrous, will also influence light absorbance or reflectivity, and must also be portrayed in an illustration. In the next drawing, we will try not only to use the proper shadowing techniques, but to practice obtaining desirable tone as well. Interestingly, the same techniques as those of shadowing are used for tone and texture. Thus, in a drawing, stipple marks can represent shadow, tone, or texture, and structure. The skill of the artist determines whether or not this interpretation becomes confusing.

Drawing 22

Obtain both a cork and a rubber stopper, preferably of nearly the same size. Prop these objects up on clay so they are always viewed in the same position. If possible, adjust your light source to make certain it comes from the upper left.

Make a pencil outline of the cork stopper on a drawing board, about 1 1/2 inches high (that would be about twice as large as a medium-sized cork stopper). Complete the outline in ink. Using tracing paper, trace several outlines of the inked cork figure. (See Figure 3.3.) You may also try completing one out-line drawing using only stipples. Practice various techniques of shading while establishing tone and texture in the cork drawing.

On one figure, add irregular short pen lines to the drawing, to simulate the pits and grain in the cork texture, before you add any shadowing. This may complicate the shading, but will realistically show both texture and shadow, and will enhance the appearance of the drawing. (See Figure 3.3A.)

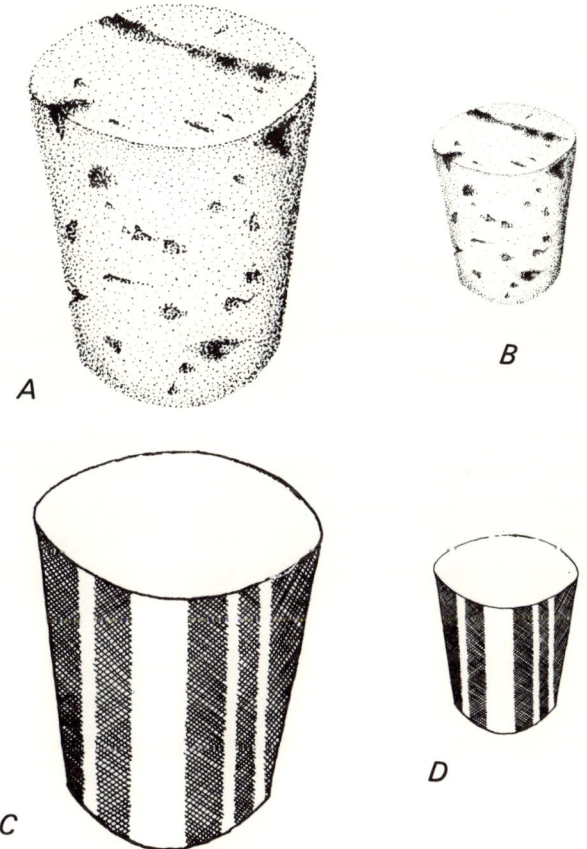

Fig. 3.3. Drawings of cork and rubber stoppers. A and C are reduced to 75 percent of the original drawings; B and D to 50 percent. Which size looks better?

Drawing 23

Follow the steps in Drawing 22, using a rubber stopper.
(See Figure 3.3C.)

Note that the black, smooth surface presents very different problems of shadow and highlights from that of cork. You may find that relatively sharp gradations of shadow and highlights are effective here. In fact, if the object were made of smooth black marble (a highly reflective surface), a sharp white highlight line would be very effective to portray this reflective texture. (Compare the highlighting techniques in 3.2C and 3.2D.)

Self-critique of Exercise 1:

1. Does the whiteness of the egg come through in your drawing, or is it your familiarity with the object that permits this interpretation? Did you use secondary highlights?
2. Which of your figures in Drawing 19 had the most desirable shadow effect? Why?
3. Which shading technique do you feel you could most easily master? Why?
4. Do the stipple lines on your drawing of the cork stopper approach the "texture" of cork? In other words, do you see both texture and tone as well as shadow in this drawing?
5. Are the ellipses portraying the top and bottom of the stoppers drawn with sufficient care (regular and apparently parallel)?
6. Do you have any line width errors?
7. Are your stipple dots round and regular, or can you detect lazy stippling?

TIPS

1. Light source
    It is standard practice in biological illustration to assume the light source illuminating the object comes from the upper left side of the specimen. This primary light source determines the position of highlights and the primary and secondary shadows. Using this positioning, a simple C-shaped arc, or

incomplete circle, would be interpreted as a pit, depression, or crater on a flat surface, since the primary shadow would be on the left (inner) side of the pit. Strong highlights drawn on the right of the pit, where the line that would have completed a full circle is absent, obscure any details of the right lip of the pit.

Had the C-shaped arc been reversed, putting the open end of the circle toward the left (or better, the upper left), the observer might interpret the object as a disc-like elevation from the otherwise flat surface. This interpretation is caused by the understanding that the upper-left light source would cast shadows to the right of the object, with primary highlights on the left. A quick scan of magazines and periodicals with many drawings and photographs will demonstrate the common use of this upper-left lighting convention in areas other than biology.

2.   Squinting to determine shadowing

It is sometimes helpful, in attempting to determine the tone of a biological subject, to view it through half-closed eyes. A squint eliminates some of the fine details and color variations, thus reducing the object to areas of relative darkness and lightness. You might also determine with half-closed eyes what degree of lightness or darkness different gray areas or shadows will need. This comparison will help to determine the amount of stippling to use to represent these gray areas on the drawing.

3.   Shading for halftone or linecut reproductions

A halftone illustration, which is photographed through a screen so it is reproduced using dots, is more expensive to make than a linecut illustration, which just reproduces the line drawing. The halftone process is used to reproduce an original illustration, such as a pencil drawing with shading or wash drawings or photographs, in which the subject is depicted in continuous tones of gray. Shades of gray are apparent in illustrations using this form of reproduction. Offset and collotype processes are other processes used to reproduce drawings with continuous shades of white, gray, and black.

Linecut processes can be used for pen-and-ink line drawings, which do not have gray shades and rely on thick and thin lines or a clustering of stipples (dots) to produce tonal quality. The more stipple dots in an area, the darker gray it will appear in reproduction, even though the original drawing has only black ink on a white surface. Because shadowing is much easier to accomplish with lead than with ink, particularly for the beginning illustrator, drawings are generally first made in pencil with shades of gray representing shadows to show shape and/or texture. Then the drawing is inked and pen strokes (hachures) and dots (stipples) are substituted for gray tones. The skill of the biological illustrator may be judged in a finished ink drawing (reproducible by linecut processes) by whether the stipples are used to represent structure, shape, texture, and/or shadow in such a way that the viewer is not confused and can easily interpret what was intended.

4. Stippling

The size of the stipples should be determined by the pen size rather than by differences in pressure on the pen point. Small dots or stipples are preferred since their use is usually to impart a degree of darkness to the drawing without making the observer overly conscious of the stipples. The larger the stipple, the more aware the observer will be of the dots. However, large, or irregular, or differing-sized stipples are a desirable technique for depicting various textures. Rapidographs, or other rigid pens, are useful for maintaining constancy of stipple size. Usually, a regular round dot is preferred, but different textures may be desired in a drawing, and these may be produced by irregular-shaped stipples, as indicated above. Stipples may be distributed regularly on the surface, or they may be irregular (random) in position, which suggests a rougher surface texture. Thus the size, spacing, and regularity of the stippling are all important in achieving different effects.

Rapid work in stippling produces poor quality. Guard Against lazy or hurried application. Quickness can result in clusters or clumps of stipples, and these irregularities will portray unwanted textures in the drawing. The pen point should be wiped often because lint or other debris on the point will produce large, coarse, or irregular dots. Corrections in stippling may be made with opaque white (see Annotated List of Supplies) covering the mistakes or unwanted texture. Let it dry completely and restipple.

It is best to tone the entire drawing with light stippling and then rework it to the desired degree of darkness or texture. If the initial area on a drawing is made too dark, it will take many additional hours of stippling to maintain the same tone or shadow throughout the drawing. Further, other areas of the drawing may have to be darkened almost to blackness to match the tone of the completed area. For these reasons, it is highly desirable to tone and shade all areas simultaneously, rather than stippling any one region of the drawing to completion first. It is important to use as fine a stipple dot as possible. If your drawing is to be reduced, use as fine a stipple dot as possible that will still show up in the reduction.

5.   Tracing techniques and symmetry

Frequently biological illustrations are of objects that have bilateral symmetry. In some cases, such as butterflies and moths, both halves are drawn even though they are mirror images. Sometimes, however, the left half of the drawing may show the upper view of the body (dorsal view), while the right half shows the lower body surface (ventral view). Individual drawings of this type may have the central axis of the left and right half separated by a small amount of space (1/8 inch), so that the viewer will not be confused by the two views on a single outline.

There are several convenient techniques for drawing a biological specimen that has bilateral symmetry; that is, the right and left halves are identical. One of these is to use a tracing box (see Annotated List of Supplies). When half the specimen (say the left half) is drawn, a tracing can then be made; then the drawing can be flipped over and a tracing made of the half in the opposite (right) position. This can also be done by tracing against a window, if you use illustration paper thin enough to see through.

A second technique is to smudge the back of an illustration with a soft lead pencil or a piece of colored chalk. By retracing the original against a second piece of paper, you can make an outline, which may then be transferred to the other half of the original drawing. Using carbon paper is not advised because of smudging and other difficulties.

6.   "Tracing" proportions from biological specimens

A rather simple glass-tracing technique may be used to establish proportions when drawing biological specimens as large as a leaf or a crayfish. Alcohol-preserved specimens such as fish may also be traced using the following procedure: A piece of glass or other type of transparent drawing platform such as Plexiglass is placed between the specimen and the illustrator. A transparent acetate sheet (frosted acetate will also work) is taped to this surface. Two or three books on either side of a dissecting tray may serve as a glass-holding platform, and the subject placed below the glass is then traced on the acetate sheet on top of the glass.   The problem of changing views by moving your head can be avoided by viewing the subject through a pinhole placed in a thin piece of cardboard (or a hole in a plastic ruler pierced with a dissecting needle) held in a constant position above the glass by a device such as a chemistry ring. The subject can then be viewed through the pinhole and traced on the transparent sheet; thus you can easily obtain the correct proportions. Varying the height of the drawing glass will reduce or enlarge the size of the tracing. When the tracing is finished, retrace it on thin tracing or drawing paper. You can easily do this by placing the tracing on the opposite side of the drawing glass and using a light source such as a gooseneck lamp to make the glass a tracing box.

4

Stippling and the Gradation of Light_____

Humans perceive images by reflected light, usually in color. But drawings with pen and ink are made only with black lines (continuous dots) and/or dots (interrupted lines) on white paper, so they cannot be exactly like the original image. For example, in Figure 4.1 the margins of a leaf are executed with a continuous line; in Figure 4.2 with a linear sequence of stipples (dots). Of course, we do not actually see an outline as we look at a real leaf, but rather a difference of color between the leaf and the background against which it is observed. So the dotted line outline may be preferred because it "softens" the surface somewhat, particularly if the dots are small enough so that we are not conscious of them as stipples, but as a reflective surface different from the background.

An entire drawing can be made with stipples, as is done with photographs in newspapers (halftones). Halftones are composed by breaking up an image into a series of small dots by means of screen filters, with the result that apparent shades of gray are produced. The viewer is ordinarily not aware of the dots because of their small size. Ideally, the biological illustrator is trying to do the same thing. Since it is physically impossible to make such small dots (hands not steady enough, pen points not small enough, etc.), the drawing is usually made large enough that the stipples can be comfortably drawn and technically controlled, and it is then reduced so that the eye no longer notices the stipples, but sees only the intended shading.

The exercises in this chapter help develop the techniques of stippling to portray lines and shadows and lay a firmer foundation for the use of texture and tone in biological drawings.

Exercise 1 deals with flat objects, Exercise 2 with round

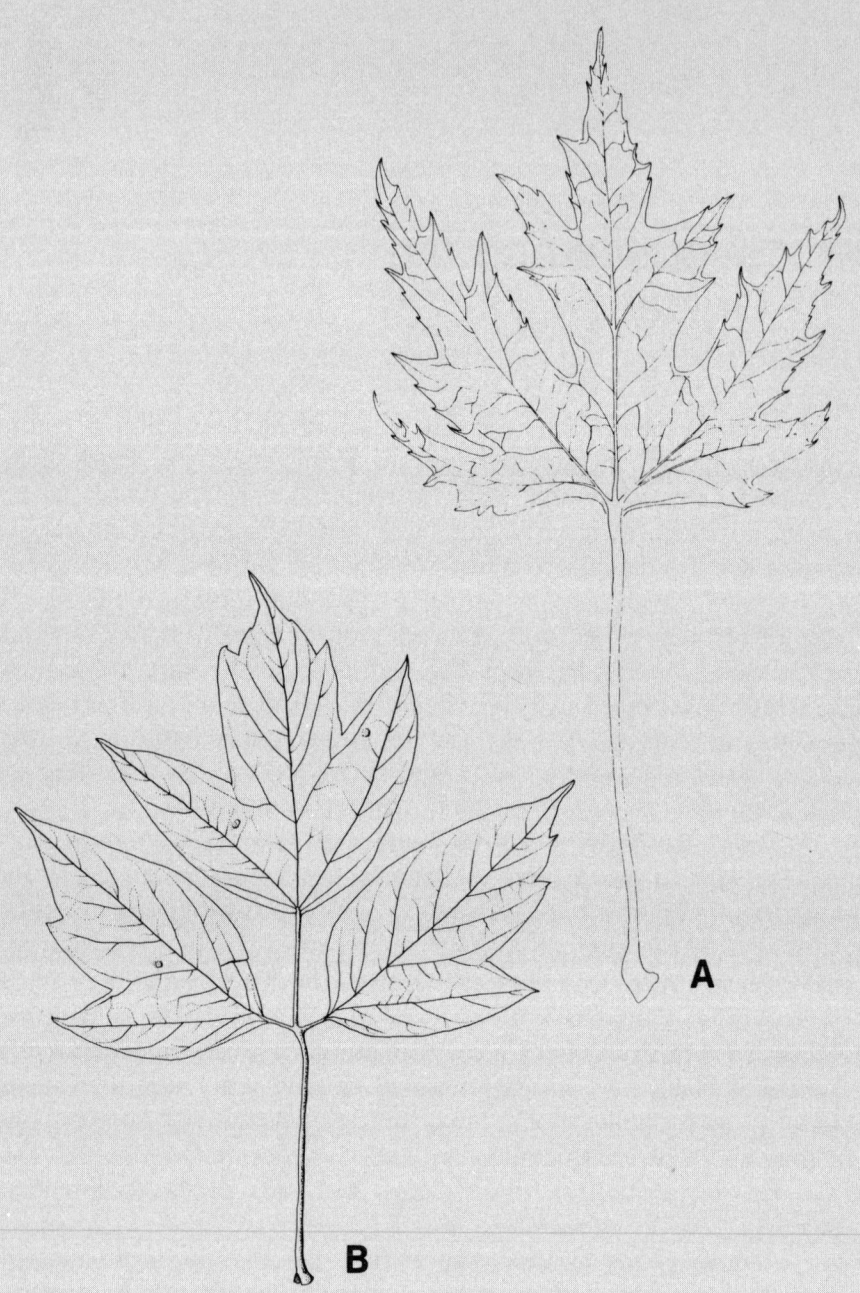

Fig. 4.1. Line drawings of leaves. See Figure 4.2 for a stipple draw-ing of the leaf in A.

objects. Exercises 3 and 4 involve the use of stipple board which is a real time-saver for the illustrator. Exercise 3 emphasizes the structure of the object, and, as skill is developed, Exercise 4 involves portraying the texture.

EXERCISE 1

Purpose: To continue developing skills with ink stippling to portray structure and tone in drawings of relatively flat biological objects.

Drawing 24

Obtain a pressed leaf and make a line drawing, with pencil, on (smooth-finish) Strathmore, showing details of the midrib and venation. Complete the drawing by tracing over pencil lines in ink. Use no shading, except perhaps a few stipples on the midrib and petiole or other prominent structures where a three-dimensional effect is desired. (See Figure 4.1A and B.)

Drawing 25

With the aid of a tracing box (see Annotated List of Supplies) or vellum tracing paper, make a second drawing of the leaf, this time using stipples to replace the solid lines. (See Figure 4.2.) (See Chapter 3, Tip 4, on appropriate size of stipples.)

In addition to the use of stipples to indicate leaf margins, leaf veins, small veinlets, or even shadows on the midrib, regular stippling may improve the tonal quality, making the figure more or less intense and giving the surface a darker or lighter tone. Too many stipples may cause unwanted texture, so exercise some care.

Mount your best efforts on Bristol board. (See Chapter 2, Tip 4.)

Drawing 26

Obtain a feather about 3 to 5 inches long and make a line drawing, first in pencil, then tracing in ink. The fluffy featherlets near the base of the shaft may best be illustrated by stippled lines. Take care to duplicate irregularities near the blade margins, where the barbs may be separated and even overlap.

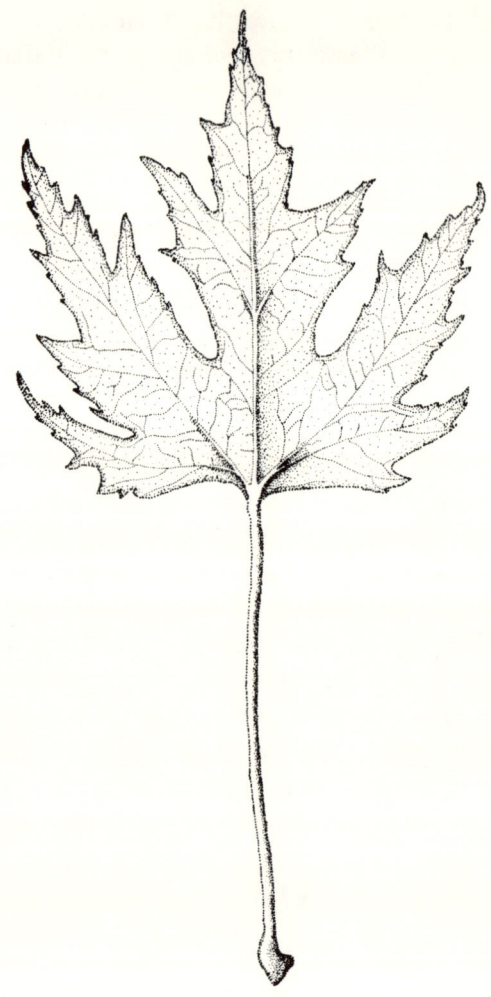

Fig. 4.2. A drawing of Figure 4.1A, using only stipples.

Complete a pen drawing, stippling as necessary to indicate details such as texture and directions of barbs. (See Figures 4.3, 4.11B, 4.11C, 4.12B.)

Self-critique of Exercise 1:

1.   Is there a distinct difference between Drawing 25 and Drawing 26?

Fig. 4.3. Ink drawings of two feathers, using stippling. The feather on the right is a reduction of the middle drawing.

2. Is structure better illustrated with line drawings or drawings done completely with stipples?
3. Is tone better displayed with line drawings or with stipples?
4. How much detail is visible in your feather? Are the barbs visible? Could you show greater detail if you made the drawing twice as large?
5. Does the feather appear heavy or light?
6. Was pigment of the feather shown in either drawing?

EXERCISE 2

Purpose: To develop techniques of stippling for texture and to continue to practice with techniques of tone in drawing relatively round biological objects.

Drawing 27

Obtain a peanut in the shell and make a rough draft drawing, in pencil, about 3 inches long. (See Figure 4.4.) The network of light lines and ridges creating a honeycomb appearance help establish the contours of the more bulbous ends of the shell. The slight elevations of these ridges and the irregular cross connections between them lend themselves to shading. Use a stump (see Annotated List of Supplies) to darken the depressions in the network and to portray primary shadows.

Fig. 4.4. A peanut drawn on fine stipple board with a pencil. Does this rendering resemble a peanut?

Drawing 28

Make a pen-and-ink tracing of the pencil drawing with a tracing box or vellum tracing paper, using ink stipples to indi-

cate shadow, tone, and texture. (See Figure 4.5A.) Develop the stippling over the entire drawing first, and gradually increase the tone to the desired intensity. (See Chapter 3, Tip 4.)

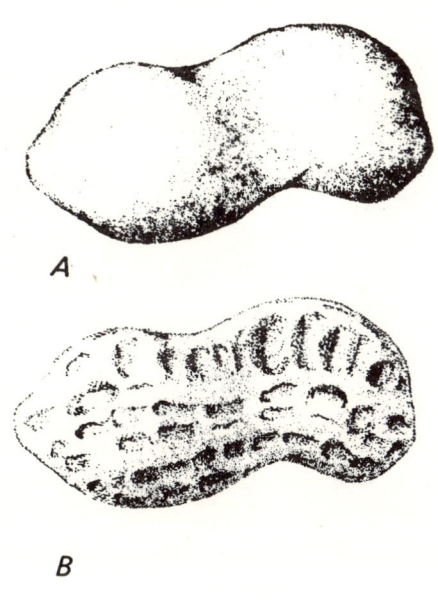

A

B

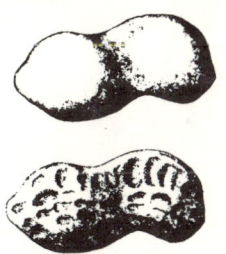

Fig. 4.5. A peanut drawn with ink stippling; the top figure shows general contours only and the surface appears to be of uniform texture; the bottom figure includes details of the irregular honeycomb network of ridges and depressions. A was reduced to 85 percent of the size of the original drawing; B was reduced to 50 percent of A. Which size looks more like a real peanut?

The biological illustrator needs to portray highlights as well as shadows. A beginning student usually masters the technique of shadowing before becoming proficient at highlighting, even though both talents seem to depend on the understanding of light sources

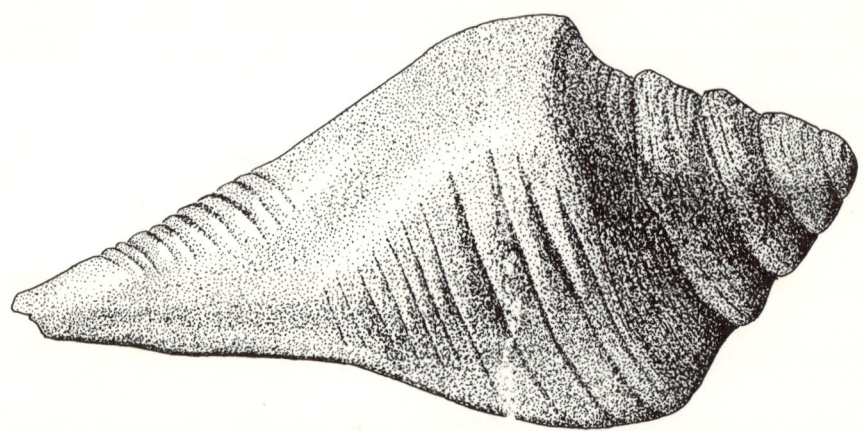

Fig. 4.6. A drawing of a shell using hand stipples on smooth paper.

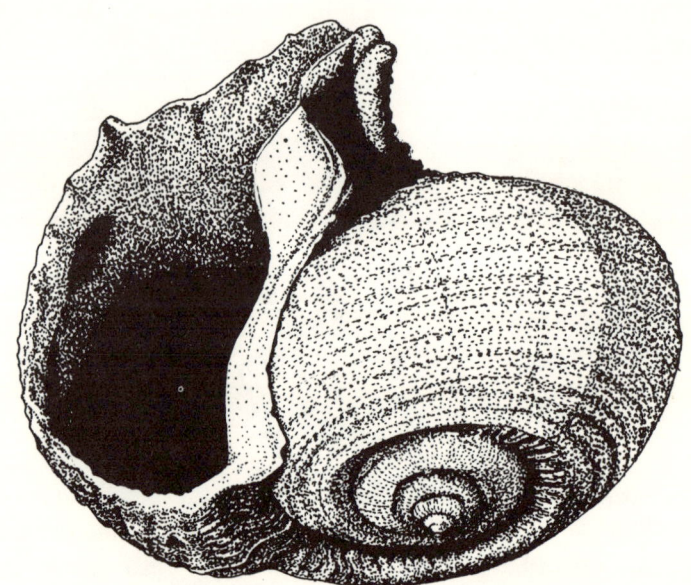

Fig. 4.7. A drawing of a shell using hand stipples on smooth paper.

and the light absorbing or reflecting qualities of various surfaces.

A number of organisms have both shiny and dull surfaces: acorns, pine cones, beetles, and vertebrate teeth. Marine shells are excellent subjects to draw when developing skills in portraying reflective surfaces together with other surface textures and tone. (See Figures 4.6, 4.7.)

Drawing 29

Obtain a highly polished marine shell less than 3 inches across. Draw the shell first in pencil on a smooth-surfaced paper using a stump for shading. Pay close attention to highlights. Copy the drawing with the aid of a tracing box on stipple board. Ink the outline.

Portraying the texture of a surface by means of stippling needs practice. Keep the following techniques in mind. The porous, light-absorbing qualities of the peanut shell, or of a cork or bark surface, are more easily drawn by the stippling technique than are the shiny, highly reflective surface of marine shells. Observing a variety of stipple drawings may prove helpful. Examine Figure 4.10A, for example; the dark, patent leather texture of this butterfly pupa is portrayed by the effective use of highlights.

Study carefully any hand-stippled biological illustration such as Figures 4.11, 4.18A, 4.18B, and 4.20. See if you can find a single professional drawing where the illustrator has used stipples to portray all these qualities: structure (segments, hairs, etc.), tone (degree of darkness, light and shadow), and texture (quality of surface structure).

Self-critique of Exercise 2:

1. Which of your drawings most closely resemble the actual subjects? Why?
2. Is the texture of the peanut visible in your drawing? Does the shell look like it would feel? Is its round nature apparent?
3. Are the highlights on the peanut drawing in balance with the primary shadows?
4. Compare and contrast your pencil and your pen drawings. Be sure to "reduce" the size of the pen illustration with a

reducing lens to view it as though it were in a published form.

5. Are the stipples coarse and irregular in some areas of your drawings? With reduction, can you note clustering, darkened areas, caused by hasty stippling?

6. Is the polished nature of your marine shell visible, that is, are the highlights effective?

Several types of rough-surfaced illustration paper (stipple board, pebble board, coquille board) are available that produce a stippled or textured effect without the effort of hand stippling. All of these have surfaces with tiny ridges or bumps separated by valleys or depressions; when a lithographic or black china marking pencil (see Annotated List of Supplies) is rubbed across the surface, a stipple effect is produced as the peaks are coated with the graphite or wax while the valleys remain white. Brisk movements of the pencil will gradually darken more of the elevations; with increased pressure, even the valleys can be coated. Some of these papers have patterns that are bold and relatively open, giving a pebbled, ridged, or grainy texture. Others have patterns that are not as definite. (See Figure 4.8.)

The choice of pattern may be determined by matching surfaces to certain textures or by personal preference. Sometimes the pos-

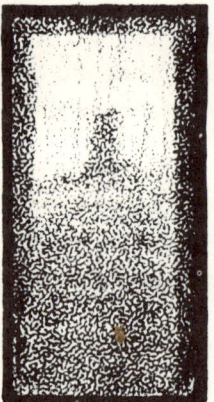 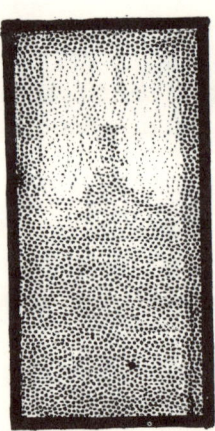 

Fig. 4.8. Examples of textured illustration board. Note differences in regularity and size of grain.

sible reduction of the size of the drawing should be considered, since the reduced texture is the one desired. (See Figure 4.9.) The finer-grained stipple boards may not reproduce well in reduction.

Fig. 4.9. A peach pit drawn on stipple board and then reduced to 75 and 50 percent of the original drawing. If you were using this drawing for publication, which size would you prefer?

EXERCISE 3

Purpose: To introduce various textured illustration surfaces.

Drawing 30

Make a tracing of the peanut in Drawing 28 on a coquille or other fine-grained textured board, using a tracing box. With a soft lead or wax pencil, draw the peanut on this textured surface and compare with the first drawing.

Take care, when using stipple board, to avoid the use of erasures. Erasing introduces smudges and affects the contours of the

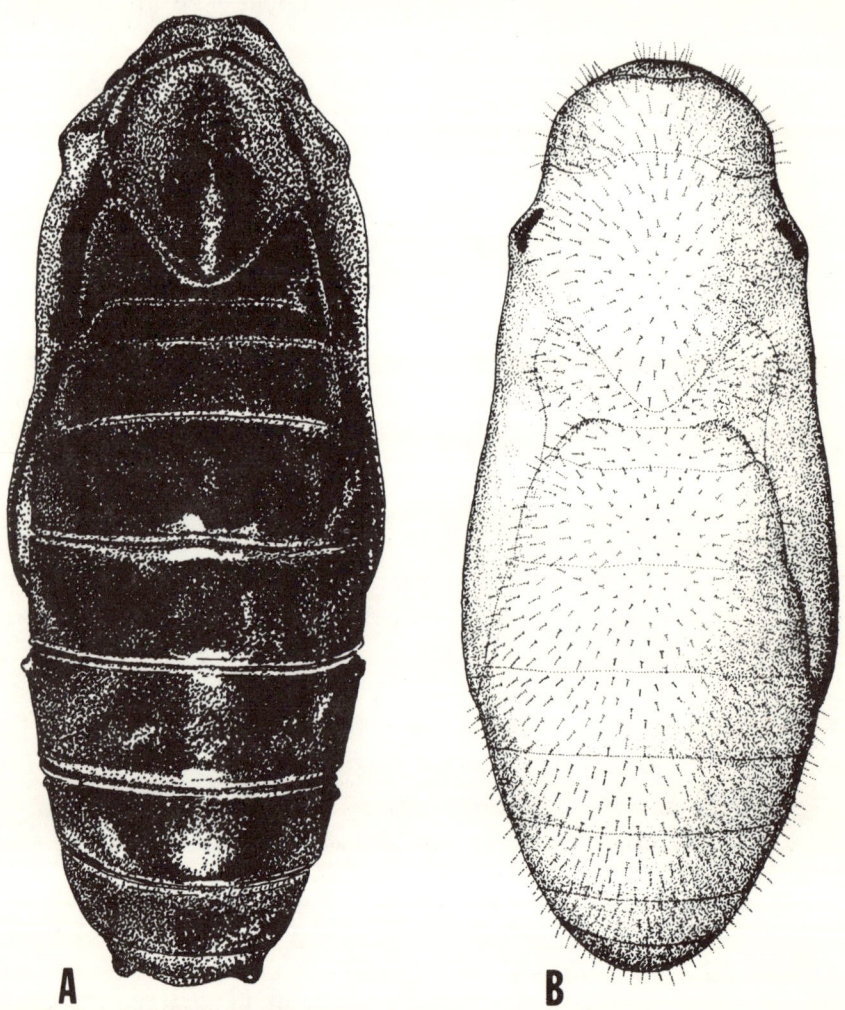

Fig. 4.10. The dorsal view of two species of butterfly pupae, drawn by different techniques. (A) Drawn by irregular stippling on glossy surfaced Strathmore. Note patent leather texture of the dark integument achieved by effective use of highlights. (Drawing time, 15 hours.) (B) Drawn by wax pencil technique on irregular-patterned stipple board, effectively portraying covering of light setae. (Drawing time, 45 minutes.)

grain of the board. Mistakes should be covered with thinned, white corrective fluid and redrawn.

Notice Figure 4.10B, which was drawn on an irregular-patterned stipple board. The outline and other body parts are sometimes drawn

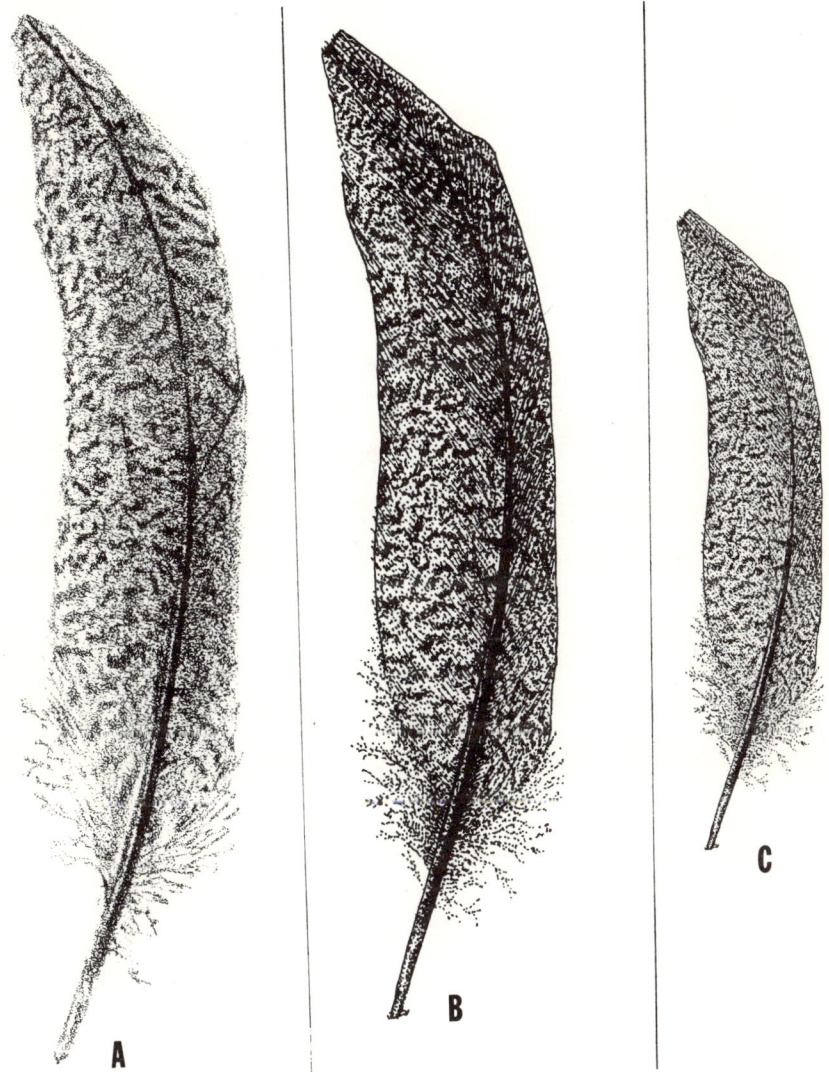

Fig. 4.11. A feather (A) drawn on coquille board, (B) drawn in ink on Strathmore (a glossy surface), and (C) Drawing B reduced to 2/3 of the original drawing. How would you improve these drawings?

with ink lines, but in this case, had the setae (body hairs) been made with solid ink lines, this butterfly pupa would have appeared to be covered with a dark, hairy coating. In this drawing, an observer first notices the shape and parts of the pupa, and only on closer inspection are the numerous setae covering the body apparent. In other words, drawing the setae with broken (stipple) lines softens the visual impact of these structures. Figure 4.11 gives the same effect with hand stippling.

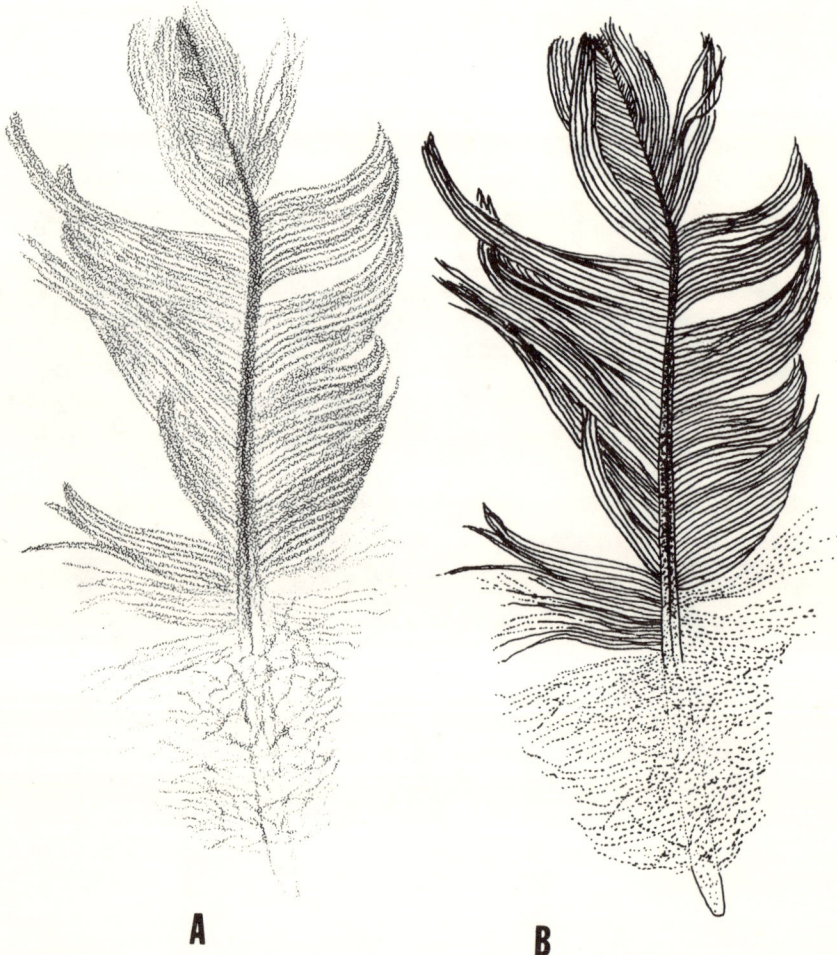

A   **B**

Fig. 4.12. The same feather drawn (A) on stipple board and (B) using a crow quill pen. (B took three times longer to draw than A.)

Drawing 3I

Make a tracing of the feather in Drawing 26 on a coquille or other fine-grained stipple board, using a tracing box. With a soft lead or wax pencil, draw the feather on this textured surface and compare with the first drawing. (See Figures 4.11, 4.12.) Try an ink outline of the feather.

Self-critique of Exercise 3:

1. Critique Drawings 30 and 31 using the same questions asked in the self-critique of Exercise 2.
2. Was the size of the grain in the illustration board you used appropriate in the peanut drawing? The feather drawing? Which drawing do you judge the best in terms of grain size?
3. Did the ink outline detract from the texture of the feather drawing?

EXERCISE 4

Purpose: To develop skills in portraying surface textures using different stipple or other textured illustration boards.

The texture and tone of an apple can often be surprisingly well captured in black and white illustrations. (See Figures 4.13, 4.14, 4.15.) The apple, like other common fruits, has distinctive features that can easily be illustrated. Pay particular attention to primary highlights, which are visible as the polish of the apple skin and can be quite reflective.

Drawing 32

Make a drawing of an apple on fine-grained stipple board. (See Figure 4.15.) Make the drawing larger than the subject itself; in other words, draw it as though for reproduction in a reduced size.

Compare the apples in Figures 4.13, 4.14, and 4.15. Which techniques do you like best? In each drawing, decide which size looks best.

Fig. 4.13. A line drawing of an apple, using hachures. The smaller figures reduced to 65 percent and 50 percent of the size of the original drawing.

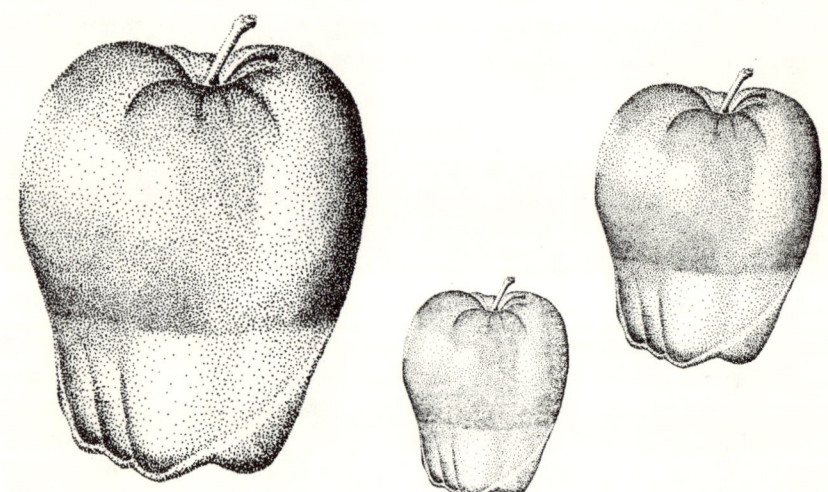

Fig. 4.14. An apple drawn with ink stippling on Strathmore (glossy paper) and reduced to 65 percent and 50 percent of the original drawing.

Fig. 4.15. An apple drawn on coquille board and reduced to 65 percent and 50 percent of the original drawing.

Citrus fruits such as the orange are easily rendered on some types of stipple board, where the grains conform to the slightly grainy texture of the rind.

Drawing 33

Draw an orange on a type of irregular-grain stipple board that you think closely matches the texture of the fruit. (See Figure 4.16, 4.17.) Determine this by pencil shading in one corner of several possible illustration boards. It will help to view the sample illustration board with a reducing lens to check the resemblance of the shaded area to the rind or peel of the orange.

An apple and orange have very different textures and tone on their external surfaces, but they do share a common roundness. Bones are another biological subject with a surface texture easy to portray with stipple board. The linear dimensions, together with the polished globular surfaces, roughened areas for connective tissue attachments, and openings for blood vessels and nerves, make them

Fig. 4.16. An orange drawn on stipple board.

Fig. 4.17. An orange drawn on stipple board and reduced to 65 percent and 50 percent of the original drawing. Does the texture of the rind come through in this drawing? Would an ink outline have helped? Is too much detail lost in the smaller figure?

more complex to draw than simple pieces of fruit. (See Figures 4.18, 4.19, 4.20.)

Drawing 34

Obtain a bone (less than 5 inches long); draw it on stipple board that matches the surface texture. (See Figure 4.18.) The

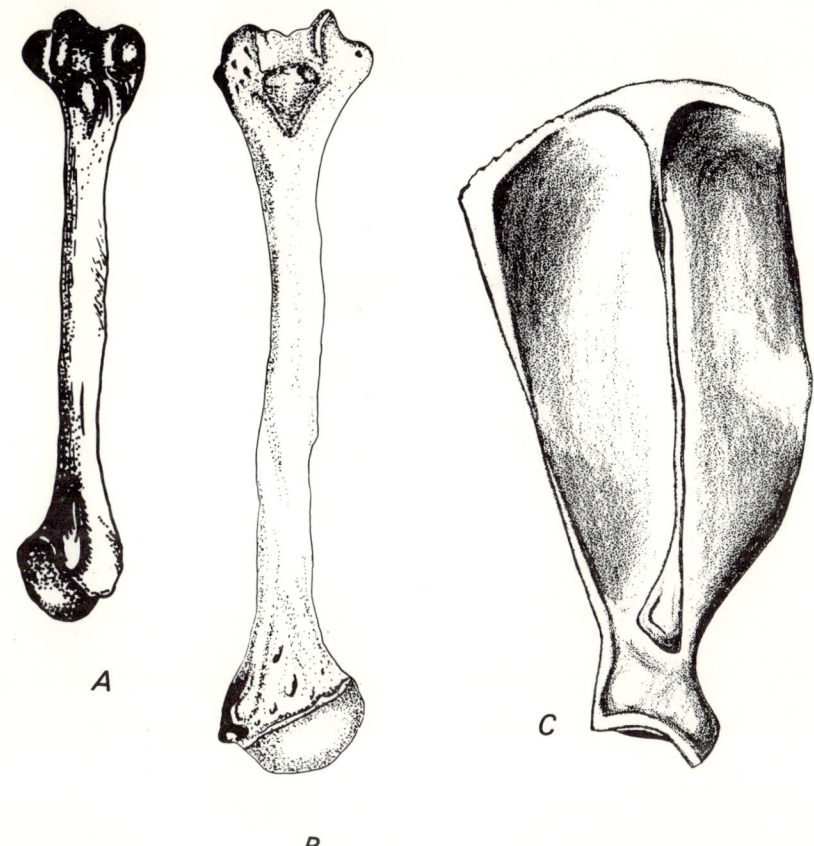

A

B

C

Fig. 4.18. Drawings of various bones. (A and B) Drawn with irregular hand stippling on glossy Strathmore; (C) drawn on stipple
board.

drawing should be larger than the subject; as you draw, keep
checking your work through a reducing lens. The outline is
generally improved by ink; highlights also improve the drawing.
You may establish proper proportions for this drawing by
using the glass-tracing technique described in Chapter 3, Tip 5.

Other examples of drawings of bone may be found in paleon-
tology books and journals. (See Figure 4.19.) If you examine these

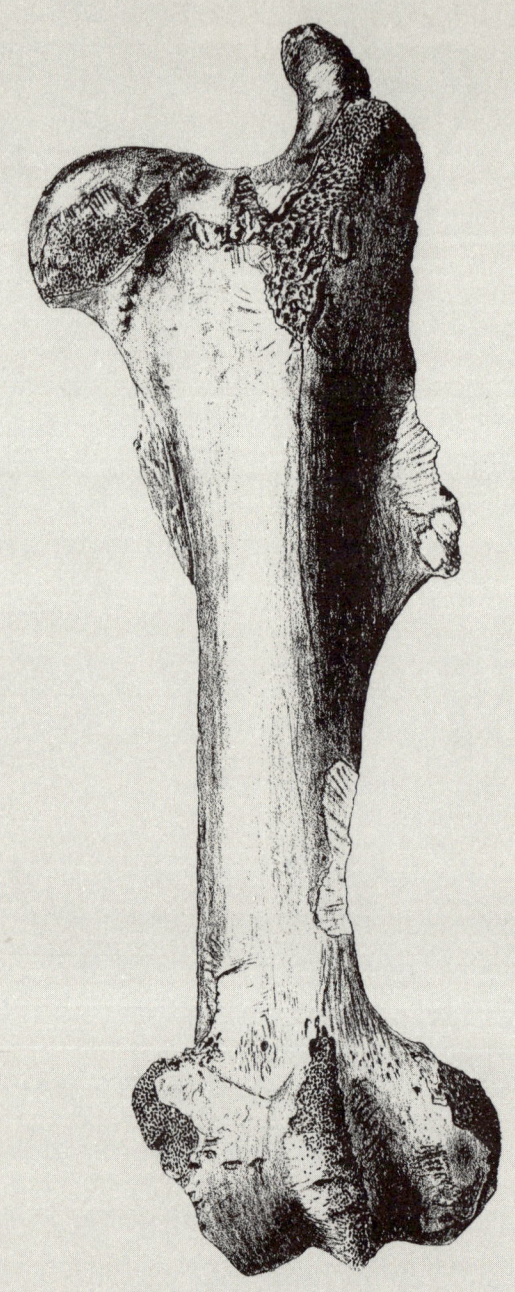

Fig. 4.19. A reduced drawing of a fossil bone. The original was drawn 18 inches long on stipple board.

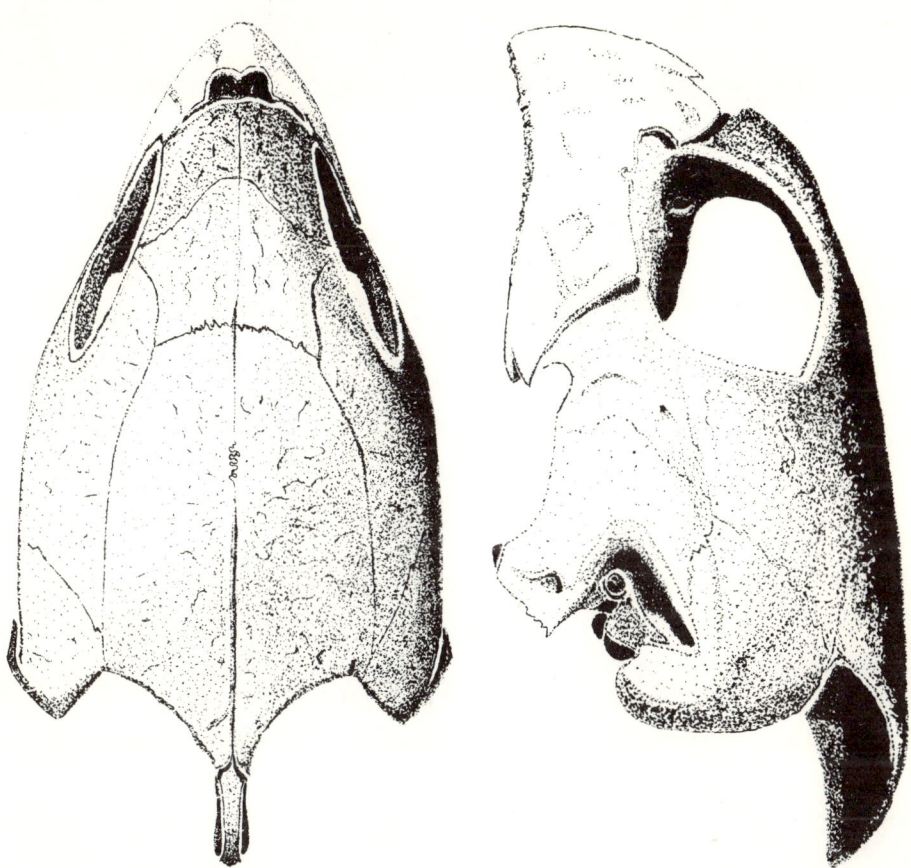

Fig. 4.20. A turtle skull drawn with irregular hand stippling on glossy Strathmore.

references, you'll find examples of several techniques. Drawings of teeth in dentistry books are also excellent examples of drawing techniques, particularly when the difference in texture of the enameled crown and the duller root area is effectively illustrated.

Self-critique of stippling exercises:

1. What was your greatest source of trouble when using stipple board? How did you try to remedy this?

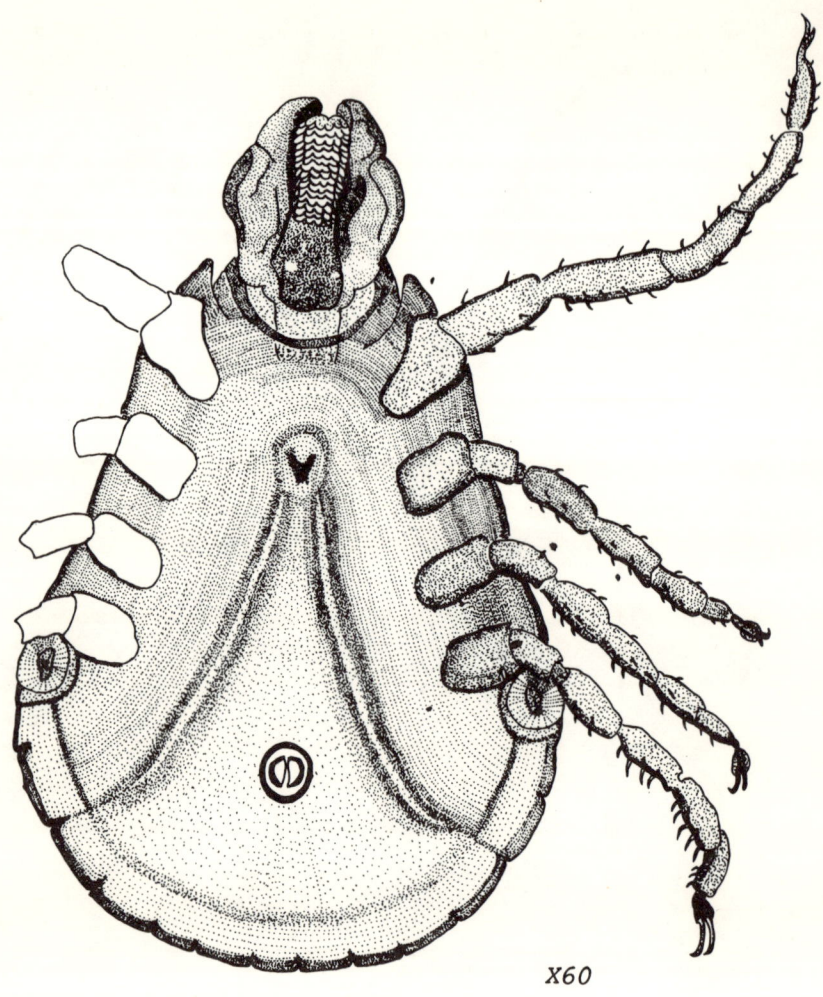

X60

Fig. 4.21. Partially complete drawing of a mite in irregular hand stippling on glossy Strathmore from camera lucida tracing. x60 indicates magnification; the scale of a drawing may be indicated in a variety of ways. (See Chapter 6, Tip 2.)

2. Which of the drawing assignments in Chapter 4 gave you the most satisfaction? Why?
3. In which drawing did you use a stipple board texture that most closely matched the actual surface texture of the subject?

4. What advantages do you see in the use of a textured illustration paper? What disadvantages?
5. Which do you prefer, stipple board or pen-and-ink hand stippling, for your illustrations?
6. Which technique is easier in terms of time and necessary skill?

TIPS

1. Highlights

The effective illustration of highlights is usually learned sometime after a beginning illustrator learns to illustrate primary shadows. But the position of the primary highlight, and even the secondary highlights and reflections in other areas on the same subject, should be considered at the same time the techniques of shadowing are attempted. The degree of reflectivity is a problem. Shiny metallic surfaces or very hard surfaces like the black plastic of a telephone usually have prominent highlights. In a drawing of these objects, the amount of light reflected tends to obscure structural details. A biological illustrator is reluctant to leave out any detail, even in the region of a highlight. And drawn with skill, a highlight can be a clue to texture and tone. A highly polished marine shell, for example, can be well illustrated with slashes of reflective highlights that lack any details at all.

Examine Figures 1.7 and 1.8 in Chapter 1. Can you find any mistakes in the illustration of the primary and secondary highlights and shadows? The beginning illustrator might be satisfied with such drawings, but more attention to highlighting details could have improved them.

Look for examples of professional drawings with highlighting. Obtain a biological or chemical supply catalog, and browse through a section on specimen bottles, flasks, and test tubes. Illustrations of glassware such as bottled drink ads in magazines often most effectively portray the highlights on these highly reflective glass surfaces. The instruments section of a biological supply house catalog has drawings of metallic tools with strong highlights, such as tweezers, scalpels, and scissors. A catalog from an art supply house may have drawings of plastic and metallic pens that also are excellent examples of effective highlighting. Books on marine shells will provide good examples of the effective use of highlights in biological illustration of the many shapes, patterns, and textures of these organisms.

2.   Uses of the blue pencil

Blue pencils are much used by illustrators. Blue does not photograph, so reduction directions or other instructions for the photographer or printer are often made in this color in the margins of drawings. Layouts and rough sketches are sometimes made in blue so there will be no need for erasing if the finished drawing is to be photographically reproduced.

3.   Drawing a three-dimensional figure: One step at a time

Analyze your subject carefully. Note the relative proportions of body parts, the length and width of structures compared with their height. Think of the subject in terms of its volume. Is it a spherical shape (like many shells)? Would it help initially to represent this shape by a circle or a series of circles as you gradually fill in the particular details? Or is the subject linear? Would it help initially to use a stick figure (much like a skeleton) to show body parts and arrangement?

Appropriate volume can then be added to the figure by geometric shapes--in the case of a human, a circle (for the head), rectangles (for the body), columns or cylinders (for legs and arms). The proportions of body parts are often drawn lightly in pencil (see Figure 4.22) and then volume is added by more and more definite lines and pencil strokes following the main contours of the body.

It is rare that beginning illustrators can draw in perfect proportion. Always use light lines to start a sketch; mistakes can be covered by additional strokes that gradually become bolder (darker) as the basic shapes and lines improve. Gradually build upon your basic lines and shapes. You may check your proportions and composition by viewing the work in a mirror; sometimes defects are much more apparent when the subject is viewed in reverse. When you are finally satisfied with your basic composition, trace the final drawing on tracing paper (see Chapter 2, Tip 3) or frosted acetate (see Annotated List of Supplies), using ink lines.

Fig. 4.22. *(A) A stick figure sometimes helps in determining body proportions and configuration. (B) Volume is then added, following the basic lines. (C) Detail is added to make a lifelike final drawing.*

5

Illustrations on Scratch Board: Try It, You'll Like It!_____

A scratch board illustration is made by cutting through a black ink surface that has been applied over a special chalk or clay-covered board. This inked surface is applied either commercially or by the illustrator. The entire surface may be treated, but in ordinary use only appropriate areas on the scratch board are blackened. The white (unblackened) scratch board may be used as an ordinary drawing surface and black ink lines applied. In doing this, however, the pen nibs tend to become fouled with the chalk particles, and continuous care and extra wipings of the pen points are advised.

We have found that the many advantages of using scratch board for drawings of biological subjects, especially those with hair and feathers, far outweigh its higher cost. With it, even relatively inexperienced students can achieve success, such as the student illustrations in this chapter, because of the ease of making corrections, additions, deletions, and changes. Using scratch board can build confidence in the beginning illustrator, encouraging more adventurous and rewarding drawing attempts.

Since the scratch board technique involves cutting through black to reveal the white beneath, the process could be viewed as focusing a light on the subject. Scratch where the object is lightest so you are adding highlights (rather than darkened shadows) in your attempt to show form and texture. Those subjects with more dark than light, or those with strong contrasts such as dark mollusk shells with their high luster or hard-shelled insects (see Figure 5.1) are especially appropriate for scratch board illustrating.

Fig. 5.1. A scratch board illustration of a beetle.

EXERCISE 1
Purpose. To develop scratch board techniques.

Drawing 35
    Obtain a piece of white scratch board. With a small camel hair brush, apply India ink to several squares (approximately 1 inch x 1 inch, in which you can practice various strokes with different scratching tools. Read carefully the tips in this chapter for the methods of applying ink and how to care for scratch board to prevent damage. When the ink is thoroughly dry, scratch the surface with the various scratch board tools described in Tip 1, including a sharp penknife or an X-Acto knife.

Practice scratching lines, grids, circles, repeated strokes, and designs, as you did with pen and ink in Chapter 1. (See Figure 1.15.)

Explore as many possible strokes as you can think of with the different tools at your disposal. Do you have a tool for scratching several parallel lines at the same time?

Drawing 36

On a sheet of drawing paper, make a pencil drawing of a small section of tree limb or trunk, or a piece of wood or bark. Shade it with a stump until you are fairly satisfied with the illustration. With a sharp knife, cut out a piece of scratch board large enough to duplicate this drawing. Mount the scratch board on a suitable Bristol board. (See Chapter 5, Tip 4.)

Transfer the pencil drawing to the mounted scratch board, and with a small camel hair brush apply India ink to the areas within the outline that will require scratching (the penciled areas in the drawing). You may prefer to ink in the original outline with an ordinary pen point, if not all the wood surface will be covered with ink. After the ink is dry (dry for 24 hours or heat over a lamp), scratch away at the drawing until you have a satisfactory duplication of the pencil sketch. (See the bark in Figures 5.3, 5.8, 5.9, and 5.11.)

One difficulty beginning students encounter in their first attempts at using scratch board is drawing "in the negative." When they are scratching white on black (rather than applying black to white surfaces) there is a tendency to whiten the shadows and leave the highlights dark. While this technique can be effectively used by the skillful illustrator, in most cases it is much better to lighten the highlighted areas, with the light coming from the upper left corner of the drawing, and leave the dark areas (shadows) alone. When drawing birds and mammals, pay attention to the reflective spots or highlights in the eyes. When eyes are illustrated properly, the entire drawing looks more professional.

Examine Figure 5.2. Some biological specimens with strong highlighted areas, such as this highly sculptured butterfly egg, can be much more easily illustrated with scratch board than with conventional pen-and-ink methods.

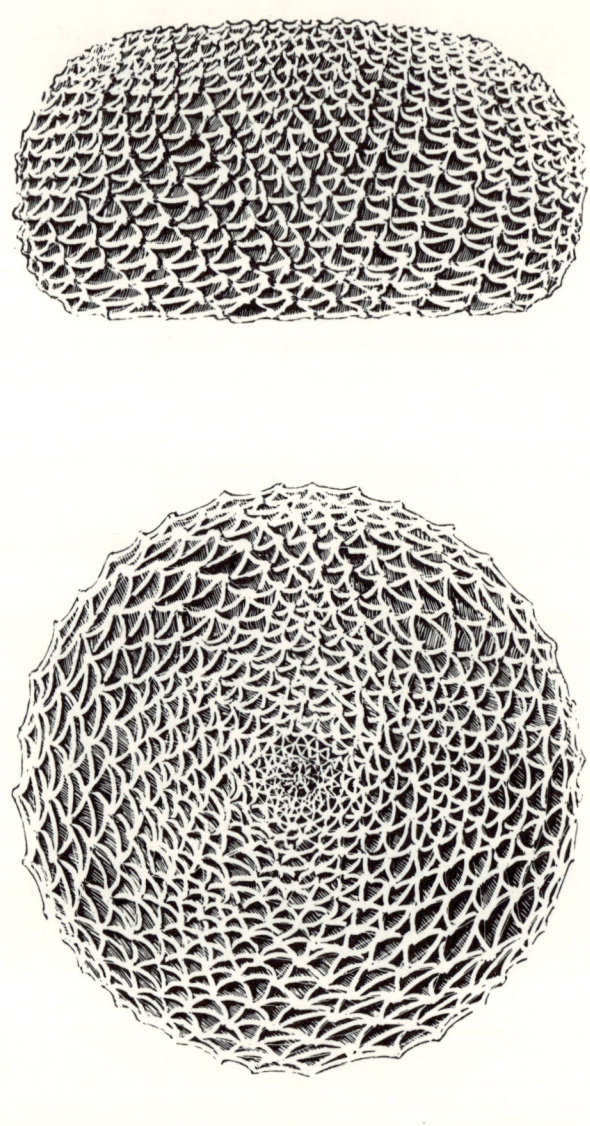

Fig. 5.2. A scratch board illustration of the upper and side views of a highly sculptured butterfly egg.

Drawing 37

Obtain a drawing or photograph (about 6 inches high) of a mammal, in which detail is apparent; a fairly simple subject is best for this drawing. Trace the subject and then transfer the tracing to white scratch board. Outline carefully those areas you need to blacken with India ink. Areas with ink are those with light contrast, of course, and in those areas the regular details may be drawn with pen and ink. After brushing on the ink and letting it dry thoroughly you may pencil outlines on the ink surface to serve as guides for scratching. Complete the illustration, referring to Figures 5.3 through 5.7 for examples of scratch board drawings of mammals.

Fig. 5.3. A scratch board illustration of a squirrel.

When drawing a mammal, it enhances the outline of the fur to place the furry texture on the body outlines with a crow quill pen, using the same strokes that will later be used with scratch board tools for the fur within the body outline. As the fur within the outline (an area blackened with ink) is scratched away, the break between the two areas can be effectively blended. Follow the contours of the body and the direction of the fur in placing the fur scratches. Use the crow quill pen for black or white whiskers.

Pay particular attention to the eye(s) in your scratch board drawing. You may wish to practice drawing the eye on spare pieces of scratch board before attempting it in your drawing.

Fig. 5.4. A scratch board illustration of a badger.

Fig. 5.5. A scratch board illustration of a panda. Notice the highlights on the eyes and footpads.

Fig. 5.6. A scratch board illustration of a fox.

Fig. 5.7. A scratch board illustration of a bison.

Drawing 38

Make a preliminary pencil drawing of a mounted or a live bird. Transfer your drawing to a mounted (or whole) piece of blackened scratch board. If the drawing is on thin paper such as vellum, it is possible to attach the paper to the top of the scratch board and, with a sharp penknife, cut directly through the pencil drawing to scratch the outline on the black surface beneath. The major outlines and proportions may be transferred in this manner, and details can then be added to the scratch board. See Figures 5.8 through 5.13 for examples of birds illustrated on scratch board.

Remember that scratch board techniques may be coupled with other pen strokes and stipple effects. The fox drawing (Figure 5.6) has crow quill hairs drawn along the body outlines, even though most of the hairs in the main body region were scratched on with tools. The bumblebee (Figure 5.14) has pen-and-ink stippled wing veins and texture; the hair outlines were achieved with simple pen strokes. You should recognize, then, the versatility of scratch board.

Do not be impatient to complete your scratch board drawing. Sometimes students who will spend 5 to 10 hours stippling an ilustration are reluctant to spend half that time with scratch board. However, experts frequently spend more than 50 hours on a single wildlife illustration on scratch board.

Self-critique of Chapter 5:

1. Did you like this method of illustrating? Was it easier for you than other techniques? Why or why not?
2. What difficulties did you encounter with scratch board?
3. Were your highlights done properly, or did you tend to scratch "in the negative"?
4. How successful were you in portraying fur?
5. Were you able to create a lifelike eye? How could it be improved?

Fig. 5.8. A scratch board illustration of a female cardinal.

Fig. 5.9.   A scratch board illustration of a blue jay. Critique the appearance of the tree bark.

Fig. 5.10.   *A scratch board illustration of a hummingbird, reduced to 75 percent and 50 percent of the original drawing.*

Fig. 5.11. A scratch board illustration of an immature owl.

Fig. 5.12. A scratch board illustration of a Japanese quail.

Fig. 5.13. A scratch board illustration of a Japanese quail.

Fig. 5.14. A scratch board illustration of a bumblebee.

TIPS

1. Scratch board tools

    Almost any sharp instrument may be used as a scratch board tool. A very sharp wedge-shaped knife (X-Acto) is good for lines and stippling, and its relatively long cutting edge permits some variation in width of scratches. A dissecting needle, a nail, and scrap pieces of metal make useful tools. Commercially made scratch board points of several designs may be obtained from artist supply houses. They resemble pen points for standard pen holders, but are solid metal, without ink slots or double nibs.

Even worn-out tweezers with points held shut by tape and filed to an appropriate scraping surface produce a satisfactory effect. Try each instrument with several different techniques-- combinations of lines, stipples, flicks, and scratches--to discover the possible effects.

2. Scratch board care

Scratch board is easily damaged. Too much humidity may lead to differences in cutting through the surface. Fingerprints, ink puddles or overlaps, and splotches of water or other oil or water marks affect the absorptive qualities of the chalking substance on the surface of the board. Bending the board, even slightly, or dropping it may cause tiny, eggshell-like cracks in the surface that affect both absorption and cutting effects. It will pay to take extra care of scratch board prior to use.

If moisture has been picked up by the scratch board, it can be dried out in a 200 degree oven for 15 to 30 minutes, or until the chalky surface is bone dry. A hot lamp may be used to dry small moisture droplets from the surface.

It is helpful periodically to blow the scrapings off the surface of the scratch board as it is being worked. Small particles of the plaster or clay may put unwanted scratches in the surface or cause undue thickening of a scratch as it is being cut.

3. Application of ink

In applying ink to the surface, several techniques may be used. Commonly a sable brush is used, but a 2-inch or 3-inch household brush is better when the entire surface is to be darkened. Air brushes may be used, or applicators of cloth, sponge, or burlap. Take care to apply ink evenly without puddling and with a minimum of overlapping, so that the covering is uniform. Use an almost dry brush; the chalk surface absorbs readily, so some practice is necessary for best coverage.

4. Mounting scratch board

Mount the scratch board to an illustration board backing with dry mount (or rubber cement, with care) before any scratching on the illustration is attempted. This may avoid eggshell wrinkles and other malformations on the finished product if you then try to cut it for mounting. Always trim with a razor or sharp X-Acto knife; scissors or cutting boards may damage the surface in the region of the cut.

6

The Use of Overlays_____

An overlay is a transparent or translucent sheet placed over a drawing for one of several reasons: to trace the illustration or part of it; to protect the illustration from dirt, smudges, or rubbing; to provide instructions to an author, editor, or printer concerning publication of the drawing; or to add printed designs or characters to the illustration. It is this last use that you will practice in this chapter.

A wide variety of art and photographic supply offices sell transparent acetate sheets with the back side coated with a waxy, adhesive layer. With a little burnishing (rubbing), the sheet will adhere to most papers and illustration boards. The overlay may contain one of a wide variety of lines, dots, symbols, designs, or letters and numbers. (See Figure 6.1.) With it, it is possible to apply patterns to drawn figures (as in Figures 2.4 and 2.5), or to fill in portions of graphs, maps, or other illustrations with patterns, to clearly define zones or areas. Letters and numbers may also be bonded to drawings and figures.

In use, the entire overlay sheet is gently laid over the otherwise completed drawing, map, or graph, and the area desired is lightly rubbed in the middle to tack it to the surface. A knife, razor, or cutting instrument (even a dissecting needle) is then used to cut along the outline of this area, and the surplus sheet is then lifted off and laid aside. Brisk burnishing with a rubbing tool or smooth instrument on the loosely attached cut out permanently bonds the surface sheet to the drawing.

Overlay sheets are fairly expensive, so take care not to allow the adhesive to get too hot or otherwise damage its sticky properties. It is wise to store overlay sheets carefully between use, keeping heavy

Fig. 6.1. Examples of some patterns in transfer overlays.

objects from pressing on them, and to keep in place the nonadhesive waxy backup sheet usually furnished with the original sheet.

The following supplies will be necessary for this exercise: an 8 x 11 1/2 inch outline map of your state or of the United States (various unlettered maps are usually available at modest cost from stationery stores or art supply dealers); at least one patterned (small dots or fine parallel lines) acetate overlay sheet to be used on the map; and a sheet of transfer or rub-on lettering (see Annotated List of Supplies, Lettering) with letters and numbers no larger than 1/4 inch.

EXERCISE 1

Purpose: To become familiar with the use of patterned overlays on maps and other illustrations.

Drawing 39

Obtain an outline map and sheet of acetate overlay patterned either with small dots or with fine parallel lines. Decide on the type of data you wish to show on the map, such as the distribution of collected specimens of a particular species. You could indicate the counties in the state where the major universities and colleges are located, or show county population levels with a key to the various patterns. Notice how the counties shaded by the acetate overlay are easily distinguished from other counties. (See Figure 6.2.)

Place your overlay pattern on top of the map and cut out the necessary section(s). Carefully remove the extra overlay material. Plan your use of this expensive material; instead of cutting out a small county from the center of the overlay sheet, use the pattern at the side or ends of the sheet to conserve as large a piece as possible for use on other drawings. After the excess is removed, carefully rub the overlay material over the entire surface so that it adheres tightly to the drawing.

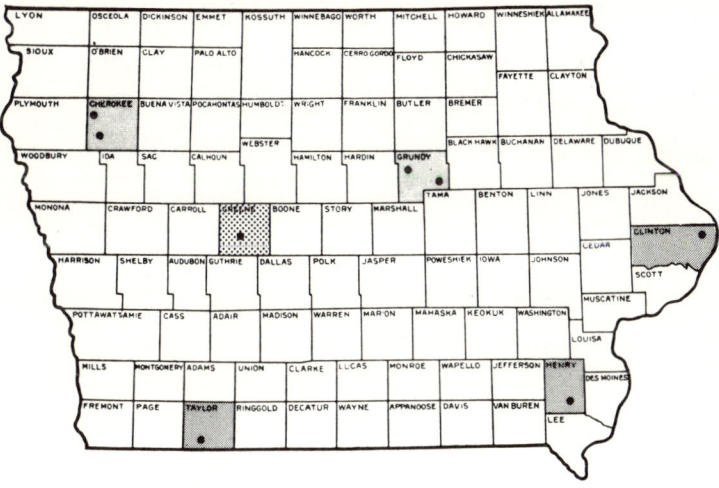

Fig. 6.2. A map of Iowa, using acetate sheet overlays to single out several counties. Do some patterns work better than others?

When using two or more patterns of overlays for different areas or values on one drawing, don't forget that the patterns may look quite different if the figure is reduced. If the drawing is to be reduced, check the final appearance by looking through a reducing lens, or if a lens is not available, hold the overlay pattern at arm's length (or squint at it) to see if you can tell the reduced patterns apart, or whether an undesirable value is introduced (such as too dark or too light a pattern).

In Figure 6.2, note that a county name was obscured by an overlay of large dots. This problem can be avoided by cutting the overlay pattern away over labeling, or by selecting a pattern that would not interfere with reading the printing.

EXERCISE 2
Purpose:To practice the use of labeling with rub-on letters.

Acetate overlays can also be used for labeling your illustration. A number of lettering styles and types are available and are an inexpensive substitute for mechanical lettering. (See Annotated List of Supplies, Lettering.) Lettering comes in uppercase (capital) or lowercase styles. The size and thickness of lettering chosen is determined by the purpose of the drawing. If the illustration is going to be reduced, clarity in the reduced version is the decisive factor.

Drawing 40
Make a simple line drawing of an irregular shaped amoeba about three inches across. The pseudopodia, or cytoplasmic body extensions, can be in any position. Add a round nucleus somewhere in the cytoplasm. Using overlay rub-on type letters, label the following structures: cell membrane, nucleus, cytoplasm, and pseudopodium. (See Figure 6.3.)

Plan the labels along with the drawing for artistic balance and design. Place the labels close to the subject. They can be flush left (all labels on the left with the first letters aligned vertically), or flush right (all labels on the right with last letters aligned vertically), or they may all start or end the same distance from the subject. Guidelines from label to body part should be horizontal, close to

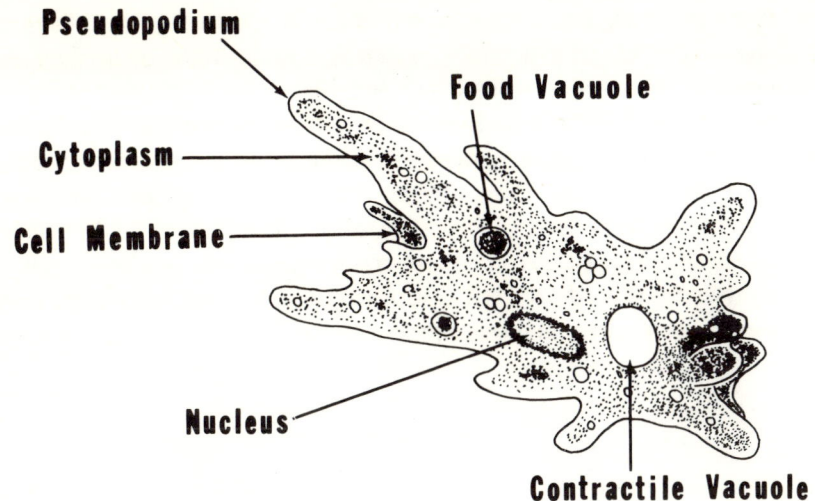

Fig. 6.3. *An amoeba illustrated by hand stippling but using transfer letters to identify various anatomical parts.*

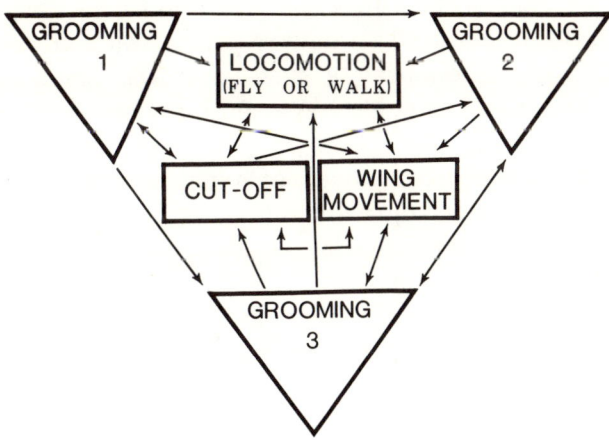

Fig. 6.4. *Rub-on letters, lines, and symbols give a diagram a finished and professional appearance.*

the label, and may be solid or broken (dotted). While guidelines may be drawn with a lettering pen and straightedge, lines or arrows are also available as overlays. (See Chapter 6, Tip 1.)

Avoid hand lettering, if possible. Drawings can be ruined by poor lettering, while sharp, neat lettering often adds to the overall effect of the drawing. Diagrams as well as drawings can be made using rub-on letters (see Figure 6.4) and overlays such as lines and symbols. To prevent peeling off or cracking caused by bending, it is best to use them on fairly stiff mounting boards or on drawings that have been attached to stiff backing.

TIPS

1.   Labeling drawings

Excellent drawings may sometimes be spoiled by poor or improper labeling. Leaving a drawing unlabeled is better than adding one or two crudely done labels to an illustration that may have taken hours to prepare. The most important rule for labels is to achieve clarity and neatness: appropriate letter size, not crowding labels, easily read and understood wording and placement, and balance and proportion between labels and figures. There are many lettering devices (see Annotated List of Supplies) that ensure a printlike label quality. Paste-on or rub-on letters from acetate sheet overlays, however, are even simpler to use. Black, black edged with white, or white arrows are also available in rub-on acetate sheet overlays. White edging on black arrows makes guidelines stand out from a drawing, even when they cross over black or heavily stippled areas on the drawings. While inked guidelines might also be drawn with a straightedge and an adjustable ruling pen (see Annotated List of Supplies), the acetate overlay arrowed guidelines are generally easier and more adaptable.

When several simple drawings are cemented to a larger composite plate, letters or numbers may be required to identify each figure in an accompanying legend. For the amateur illustrator, a convenient source of a small quantity of such numbers or figures are used calendar pages. These numbers or figures may be rubber cemented to the plate.

2. Indicating scale

The size of the subject of a drawing may be indicated on the illustration by a number of techniques: a line drawn next to the subject indicating the actual size (taking into account any later enlargement or reduction of the drawing on duplication); a segment of a line next to the subject labeled by a unit of measurement corresponding to the specimen size; labeling the subject by times magnified or times reduced directly on the drawing (that is, 1/3 size, or x 400, times meaning 400 times larger than life); or indicating the size of the subject in a legend (caption). Because of the problems of size with reproduction, it is usually more convenient to use a short line, labeled with a suitable standard in the metric system, to indicate the scale of the actual subject.

Acetate sheet overlay. See also Artype; Craftint; Frosted acetate; Plastic film; Zipatone; Chapter 6, Tip 1.

A transparent sheet, available in many types, designed to protect drawings, to trace them (see Frosted acetate), or to shade or pattern various portions of an illustration. Acetate sheets made for shading or patterning contain lines, dots, printed designs, or symbols that can be transferred to the drawing. Such overlays usually have waxy adhesive backing so that they can be rubbed on (burnished) to any illustration and appear as part of the original drawing. They are cut out from the original sheet laid over the drawing; the excess acetate is then removed, leaving the design in place.

The types of patterns, lines, stipples, letters, and special figures available on acetate sheets are published in catalogs of art supply dealers.

Adhesives. See also Dry mount; Rubber cement; Chapter 2, Tip 4.

A number of products manufactured to bond surfaces together. These include two-sided tapes (cellophane tapes with adhesive on both sides), wax coatings (dry mount), and various liquids (glues, rubber cement, etc.). For most biological illustration work, rubber cement is adequate and often preferred.

Art gum eraser. See Erasers.

Artype. See also Lettering.

A brand name of a type of overlay that can be transferred from overlay sheet to illustration by rubbing. The term has become generally used because this brand was one of the original, widely used transfer types.

Bristol board. See also Illustration board; Papers.

A heavy, opaque, drawing paper (or board, if cardboard backed) available in two surface textures: kid (nonreflective, matte finish) or plate (smooth, glossy finish). Bristols vary in thickness or weight, from 1 ply (thinnest) to 3 ply (thickest).

Brushes. See also Dust brush.

Small paint brushes are used to add wash to pen drawings or to apply ink to scratch board. Artists use fine, red, sable hair brushes or camel hair brushes (numbers 2, 4, and 6 are basic); but small, inexpensive, dime store brushes may be used instead. Either flat or round brushes are satisfactory, but the flat brushes are better for applying wash.

Small hair brushes are also used to apply white gouache paint or other corrective liquid materials to drawings. Thin applications are preferred since upon drying the paint should be sufficiently hard to accept the ink corrections to be placed on top without the ink lines feathering or running.

Burnishing tool

A device for rubbing letters or patterned overlays on to drawings. They may be small rounded knobs on the end of a penlike device, or flat, smooth, (bone, wood, or plastic) devices with a wider rubbing surface that diffuses the applied pressure to a larger area of the overlay.

Camera lucida

A device with a lens used to reduce, enlarge, or trace an image projected from a source to a copy paper. In biology a camera lucida attachment on a monocular microscope makes both the object and the drawing pencil visible on an illustration board as a superimposed image. If the head (or eye) is not moved, one can make a fairly accurate tracing of a slide specimen.

Carborundum

A synthetic abrasive stone used in biology labs for sharpening scalpels, tweezers, and scissors. It may also be used to sharpen, polish, and clean drawing instruments, pen points, and cutting knives. Whetstones (or oilstones) are carborundums on which a drop of moisture, usually oil, is applied during the grinding stage.

Cement. See Adhesives; Rubber cement.

Chamois skin

A soft, durable, highly absorbent skin used for wiping pens and cleaning drawing instruments. The thickness of the material keeps ink from soaking through to the fingers. Lintless rags or even tissues may be used for wiping pens and pen points, but one must be careful not to pick up lint on the pen nibs.

China marking pencil. See also Lithographic pencil or crayon.

A greasy or waxy pencil, often used with stipple boards. They are often paper-wrapped so that the waxy center can be exposed by removing layers of the scored, spirally arranged paper covering. They may be additionally sharpened with a razor as needed.

Compass

A mechanical instrument for making circles or arcs of circles. There are many sizes and types, with separate pen and pencil attachments. In use, take care not to scratch the surface of the illustration paper with the point, making a hole that can collect ink.

Copy box. See Tracing box.

Coquille paper or board. See also Papers.

A brand name of a textured type of stipple paper or board that makes easier graduations of shading. A china marking or lithographic pencil is usually used if the illustration is to be reproduced by linecut techniques. While soft pencil may also be used effectively on coquille board, the illustration could be reproduced only by a halftone process.

Corrective paint. See Brushes; Erasers.

Craftint. See also Acetate sheet overlay; Zipatone.

A brand name of a type of transparent acetate overlay with printed patterns. These various patterns, and other acetate sheet overlays such as Zipatone, are also sometimes called Ben Day screens because Day invented the use of screens to create halftone effects on printing plates.

Crow quill pen. See also Pen points; Chapter 1, Exercise 2.

A type of drawing pen point used for making extremely fine lines and details. It has a rounded base that requires a special crow quill pen holder.

Curves. See French curve.

Cutout letters

Recycled letters or numbers from old desk calendars, etc., used on drawings or composite plates when a limited amount of lettering or numbering is needed. These numbers and letters are cut out and affixed with rubber cement. The availability of transfer (rub-on) acetate sheets has lessened the use of this source of letters.

Depression microscope slide. See Reducing lens.

Drafting tape. See Masking or drafting tape.

Draftsman's pen. See Ruling pen.

Drawing board

A square or rectangular wooden surface available in a variety of sizes and woods, the harder and more durable the wood, the higher the price. Portable boards are the most useful and can easily be propped up at a desirable angle. Coupled with the use of a T square along one edge of the board, it is most helpful in aligning drawings, lettering, etc.

Dry mount

A waxy adhesive sheet, available in most photographic stores, used to affix illustrations to Bristol or other mounting board. The dry sheet is cut to the same size as the drawing, inserted between the drawing and mounting board, and heated with modest pressure to bond the two surfaces together. A warm household iron can provide the necessary heat (cover the drawing with some paper toweling before applying the iron to the surface). Commercial dry mount presses and hand welders are available.

Dust brush

A soft-bristle brush used to keep drawings free from eraser crumbs, dust, etc. The brush must be kept dry and clean so it won't mark the drawings on which it is used.

Erasers. See also Dust brush.

Materials used for cleaning and removing surface dirt and pencil lines without smearing. They should not weaken newly inked lines.

Pink Pearl is a soft, pink, rubber eraser used for removing graphite pencil marks. The rubber comes in varying degrees of hardness; the harder the eraser, the more likely it will be to dig marks into the paper or otherwise damage the illustration surface. Other pliable erasers come in a variety of sizes and many have edges beveled specifically for pencil lines.

Art gum erasers are made from a soft gum material that does not affect the hardness and texture of the drawing surface. They come in different sizes: 1 inch x 1 inch x 1 inch, 2 inches x 1 inch x 1 inch, 3 inches x 1 inch x 1 inch. The residual crumbs from erasing continue to be effective in removing smudges, as they are rubbed across the surface.

Kneaded rubber erasers can be molded to any desired shape (and also used from propping up objects) and are ideal for removing smears and smudges.

An erasing shield is an inexpensive, flexible, thin metal plate with slots of varying sizes and shapes through which erasures can be made. This shield protects other parts of the drawing during the correcting process.

Corrective paint comes in a variety of containers and under a variety of trade names. A white, quick drying gouache (opaque watercolor) is satisfactory for retouching and correcting mistakes, unwanted lines, or smudges. It can also be used for highlights on certain types of illustrations. See Chapter 1, Tip 6.

Pumice or sandstone may also be used for removing ink lines from certain types of paper. Care must be taken not to roughen the surface, or new ink lines will feather or appear uneven.

Erasing shield. See Erasers.

Fixative. See also Chapter 1, Tip 3.

A waterproof, durable liquid that protects drawings from external damage. Clear fixatives may be obtained in aerosol spray cans. They may be clear acrylic plastics, some with a sheen or gloss; or they may have a nonglossy or matte finish. The most preferred kinds are a fast-drying material with a finish that will not affect tone and texture or discolor halftones, and kinds that allow corrections or changes to be made to drawings even after protecting them with spray. Select an odorless brand, if possible; some sprays have an unpleasant smell.

To use, quickly cover the surface of the drawing with a light coat of fixative sprayed from about 12 inches away. Avoid splattering, soaking, or wetting the surface, which may cause lines to bleed. It is better to apply several thin coats rather than risk the danger of damage from one heavy soaking.

French curve

A mechanical drafting tool, usually made of plastic, containing a variety of arcs or curves that can be used as guides for pen or pencil lines.

French stump. See Stump.

Gouache corrective paint. See Brushes; Erasers.

Frosted acetate. See also Acetate sheet overlay; Plastic film.

A plastic sheet with a matte surface that, unlike shiny plastics, accepts both pen and pencil lines as well as other media. The frosting, which is only on one side, does not interfere with transparency, making this an excellent tracing medium. Errors may easily be removed with a moist towel (most inks), an eraser (pencil), or careful scraping with a razor blade (dried pigments).

Hollow point pen. See Reservoir pen.

Illustration board. See also Papers.

Any one of a number of cardboard-backed drawing papers

(Bristols) available in either kid (matte) or plate (glossy) finishes. They are usually sold in full sheets (40 inches x 60 inches), but many art dealers cut the boards to smaller sizes. The backing cardboard supporting the Bristol comes in several thicknesses (plies).

Ink
A liquid used to make a drawing permanent. Drawing inks should always be waterproof. A good quality, absolutely black India ink should be used; poor quality results in lack of a uniform dark color, and erosion of the surface with erasing. Several acceptable brands are on the market. Higgins "Black Magic" India Ink is thinner than some of the others, which means that it takes longer to dry, an advantage because ink that dries too rapidly clogs some drawing and lettering instruments. Pelikan India Ink is a popular German ink that is slightly thicker and dries more quickly. Each brand has a stopper with a dropper attachment so that ink may be conveniently applied to reservoir pens and pen points. Rapidograph pens need a particularly good grade of ink since clogging of parts can be a serious problem; Castell Higgins Ink (Tusche No. 404-80) is a good choice.

Kneaded rubber eraser. See Erasers.

Knives. See also Razor blade; X-Acto knife.
A number of different illustrating functions are served by various knife types. These include cutting illustration board, paper, overlays, and protective stencils, and sharpening points on lead and wax pencils and stumps. A single-edged razor blade is an effective substitute for the more expensive knife varieties that have only one specific function. However, a good X-Acto knife with several blades is also most useful.

Layout pencil. See Pencils.

Leroy. See Lettering.
A brand name for a number of different mechanical lettering tools and instruments.

Leroy socket holder. See also Pen.
An inexpensive attachment fitting standard pen holders,

equipped to hold reservoir pens. The socket holder is adjustable to the slope of the user's hand so that the reservoir pen is held at a right angle to the paper.

Lettering guide or template.   See also Lettering; Mechanical pens.

A device, usually plastic, that mechanically guides reservoir or fountain pens to form lettering.

Lettering.   See also Mechanical pens.

For mechanical lettering, there are several types of guide pens and sets available. Rapidesign or Rapidoguides are plastic templates designed for use with Rapidograph pens. Like all sets, the characters may be purchased as uppercase (all capitals) or as lowercase letters. Slanting or vertical letters and numbers are also available. The height of characters varies and must be considered in drawing for reproduction. Most sets are designed for use with a T square or a guide holder that keeps the template level.

Wrico lettering guides require special Wrico lettering pens of a reservoir type. They are reasonably priced, although purchasing the great number of stylets and pen sizes could make a complete set relatively costly.

Leroy lettering equipment is an excellent device that requires Leroy reservoir pens. Rather than placing the pens through a cutout stylet to print, the Leroy uses a scriber (special tracing pen) that traces the letters from a template to the paper surface.

A relatively inexpensive substitute for mechanical printing is transfer-type (cutout, paste-on, or rub-on) letters available on acetate sheets. Individual letters are selected from a printed pattern of letters of particular styles and sizes. The Artype brand has characters on the front of the sheet; each letter is cut out, placed in position, and rubbed so that the waxy back of the transparent sheet adheres to the surface—letter, sheet, and all. With another type, such as the Instant Lettering brand, the letters are on the reverse of the sheet, and after properly positioning a selected letter, burnishing with a ball-point pen or rounded object on the front of the sheet transfers the letter from the back of the sheet to the drawing surface. When using transfer types of lettering, proper alignment is always a problem, and letters may be distorted or crack as they are applied.

Light box. See Tracing box.

Lithographic pencil or crayon
A greasy, black wax crayon that comes in varying degrees of hardness. Variation of pressure on application will produce thicker or thinner lines, but not darker or lighter ones. They are used with pebble or stipple board to produce drawings suitable for linecut reproduction rather than the more costly halftone process.

Masking or drafting tape
Tape used to hold drawing paper to the surface. It is preferred over transparent tapes, which may damage the paper on removal, and over thumbtacks and other pinning devices, which leave pinholes. The better varieties do not leave glue residue on the paper surface.

Mechanical pens. See also Lettering; Reservoir pen; Ruling pen.
Certain categories of pens (reservoir pens, ruling pens, etc.) are described as mechanical, no doubt because of the accuracy they impart to technical drawing and also because of the special devices such as templates and stylets used with the pen to achieve a print-like quality. Many mechanical pens are much like the common fountain pen, holding their own ink supply that can be periodically refilled. They also have a single point, usually needlelike and hollow, through which the ink flows to make a line or dot of unvarying width. The points of many of these pens can be changed to get various line sizes.
One of the standard brands, with a name now almost synonymous with mechanical pen, is Rapidograph. Its tip is arranged on the point for ease in working with rulers, templates, or lettering sets. The pen must be covered and upright when not in use to prevent the ink from drying in the inner core of the point and clogging the mechanism. Many illustrators store them upright in closed, sometimes moistened, glass jars. Pen sizes range upward from 00 (the thinnest line); sizes 0, 1, 2, and 2 1/2 (thicker lines) should be adequate for most biological illustrations. Larger sizes are available for special purposes.
Other brands are also excellent, such as the Castell TG technical drawing pen (with 00 point). This variety has a cap that is

equipped with a moisture pad to keep the pen from drying out during storage. Periodically the cap is moistened with a drop or two of water.

Opaque projector

An instrument for projecting the image of a solid object. This is useful, even with biological material preserved in liquids, for making outline drawings and drawing accurate proportions. The image is projected to the desired size onto a paper taped to a wall or a propped-up drawing board. A darkened room may be necessary as the projector only works with reflected light.

Opaque white. See Erasers.

Overlay. See Acetate sheet overlay; Artype; Lettering; Zipatone.

Pantograph

A mechanical wood or metal device designed to enlarge or reduce artwork. In use, the pantograph is fixed in position, and while a stylet is traced over the drawing, a pencil or drawing tool on the other end of the instrument copies the action in the enlarged or reduced proportions to which it is set.

Papers. See also Bristol board; Coquille paper or board; Illustration board; Poster board.

Selection of paper is most important to achieve the most realistic and professional drawing possible. A pitted, roughened fossil bone may best be illustrated on a grainy surfaced (textured) paper; while a smooth, glossy marine shell is better drawn on a glazed or plate finish drawing paper. Generally speaking, the rougher the paper surface, the harder the pencils must be to effect a smooth tone or line. Plate or kid finishes are desirable for pen work; these finishes are smooth, with less friction between the pen point and paper surface. The nibs of a flexible point like the crow quill may be spread to produce thicker or thinner lines as desired. Matte surfaces have a dull finish better suited for types of lead and wax pencils.

Drawing paper comes in varying thicknesses or plies: 1 ply, 2 ply, 3 ply, etc. The more plies, the thicker the paper (and the more likely it is to be called a board) and the less likely the need for mounting the drawing on a supporting plate. The heavier papers are referred to as illustration board, even when they are backed by cardboard. Advantages in handling may be outweighed by their being very opaque, which makes them useless for copy work, even over a tracing box.

Following are several useful brands and types of paper:

1.   Strathmore. a brand name for a reasonably priced drawing paper. We recommend a 2 ply, glossy surface paper, which has an excellent surface for most ink drawings. It is enough to hold up under handling and to erase well. It comes in large sheets and must be cut for the greatest economy.

2.   Tracing paper: paper sufficiently translucent to permit pencil or ink tracings of drawings over which it is placed. It is frequently called onionskin or tracing vellum. The better tracing papers withstand some erasures and permit lines to be drawn clearly and sharply over the erased area.

3.   Stipple board (pebble board, coquille board): a rough-surfaced board with differing patterns of coarseness and structure, designed for use with soft pencils or wax lithographic pencils. The projecting peaks of the rough surface are the first to receive the graphite or wax material from the pencil; harder or repeated movements over the surface darken the area as the valleys (depressions) between the peaks receive more of the dark material.

4.   Scratch board (scraper board): an illustrating surface with a smooth coating of plaster or clay overlying thin cardboard. Ink is painted or drawn over the chalky surface, and by scratching the surface with a sharp tool the artist exposes the white coating beneath. It may also be purchased with the surface already black. Care must be taken not to crack, bend or chip the surface, and greasy fingerprints and moisture may also adversely affect the ink absorbing qualities of the scratch board. It is many times more expensive than other drawing boards.

5.   Ross stipple board (apparently no longer produced): a grainy textured stipple board (of various grades) overlying a thin coating of material so that it serves also as a scratch board, increasing the potential of the surface for illustrating. It is rather expensive compared with other types of drawing boards.

6.   Polyester and acetate film, glass cloth and other materials are increasingly used as drawing media instead of paper; reported to improve certain types of illustration possibilities.

Pebble board. See Papers.

Pencils.   See also China marking pencil; Lithographic pencil or crayon.

Pencil leads of varying softness give a wide variety of effects in pencil drawings. The mixture of clay and graphite determines the degree of softness. The types are identified by letters and numbers: hard (H, 2H, 3H, 4H), medium (HB), soft (2B, 4B, 6B), very soft (layout pencil).

Pen.   See also Lettering; Mechanical pens; Pen holder; Pen points; Reservoir pen; Ruling pen.

A pointed instrument that uses ink.

Pen holder. See also Pen points.

A tapered wood cylinder about pencil size that is slotted at one end so that pen points may be inserted; inexpensive. To lessen rolling, affix masking tape to the cylinder, or square off an edge of an enlarged end of the holder. Pens rolling off desks and hitting the floor with the pen point foremost has damaged many fine pen points and upset many illustrators.

Different holders for crow quill and standard pen points are necessary. The pen point should not wobble in the holder. Do not dip the holder in the ink, but apply ink to the bottom of the pen point with the ink applicator attached to the bottle stopper.

Pen points. See also pen holder.

Curved pieces of metal shaped to a point, with a slit forming the point into nibs. The shape of the point, the size of the slit, and the flexibility of the metal used in the point determine the nature of the line produced.

There are two main types of points: drawing and lettering. The drawing type is available in countless variations of points, angles, and styles which produce a variety of lines or effects. Most fit standard pen holders. One that does not, and is a must for biological illustrators, is the crow quill pen. It has a superfine point and is quite flexible.

The lettering type of point generally has an ink reservoir in the form of a hinged or flexible flap that covers the top of the nib. A drop of ink is introduced with the dropper between the flap and the nib, and the ink flows down the slot between the two surfaces as the pen is used. The pen point on lettering pens is larger than on drawing pens, and is often bent at an angle to the nib. One commercial variety is the Speedball point, which comes in various sizes and styles designed for drawing thicker lines for lettering—the higher the number, the smaller the tip; B (rounded tips), especially size B-5 and B-6, are useful in biological illustrating. A "flicker" Speedball pen features a double flicker hinge device that simplifies cleaning.

Pink pearl. See Erasers.

Plastic films. See also Acetate sheet overlay; Frosted acetate.

Thin, transparent plastic sheets generally used for overlays; purchased as a single sheet, a pad, or a roll. They may be colored, may have an adhesive backing, and may have a surface that is frosted in order to accept pencil and pen marks. Shiny plastics will usually hold only acetate or acrylic paints.

Poster board

A relatively inexpensive cardboard covered with white or light-colored paper. They are usually used for sign and display work and are not of illustration quality.

Press-on lettering. See Lettering.

Pumice. See Erasers.

Rapidesign. See Lettering.

Rapidoguide. See Lettering.

Razor blades. See also Knives.

A single-edged razor blade is a good cutting tool for illustration board and drawing paper. It can also be used to trim and erase ink lines, and sharpen lead and wax pencils and shading stumps.

Reducing lens

A concave lens that is relatively inexpensive and available at most hobby shops and photographic stores. With it, the drawing can be observed as it would look when reduced for duplicating. During the course of drawing, an illustrator should use a reducing lens several times to check on thickness of lines, whether stipples can be detected on reduction, and whether the tone of the drawing will be too light or too dark in a smaller size. A depression microscope slide may be used as a reducing lens.

Reservoir pen

An inking point or pen containing a V-shaped ink reservoir from which ink flows to the point like liquid through a hypodermic needle. The tip of the point is inflexible and produces only one thickness of line. Usually the nib point contains a fine wire around which the ink flows through the duct to the nib end. The wire can be moved to adjust the flow of ink and break up ink clogs. These points should be carefully cleaned after each use and soaked periodically in ammonia to prevent ink buildup.

Ross stipple board. See Papers.

Rubber Cement. See also Chapter 2, Tip 4.

A viscous liquid used as an adhesive to affix illustrations to protective matting or other boards. Keep the container tightly closed when not in use as the solvent quickly evaporates and the cement dries out. It may be thinned with rubber cement thinner.

To use, apply the cement quickly and evenly on both surfaces to be bonded. Allow to dry. Lightly position the drawing and firmly press or burnish the two surfaces together. Excess cement seeping out can be finger rubbed away; in the process this acts as a rubber gum eraser to pick up dirt and smudges.

Rub-on letters. See Lettering.

Rule

A straightedge tool equipped with a measuring scale. Both inch and metric rules are available. Beveled edges permit use as a straight-edge with a ruling or draftsman's pen.

Ruling pen

A mechanical device for producing lines of constant width. The two metal blades forming the nibs can be adjusted by turning a finger screw to produce from a very fine to a thick line. The blades are first positioned to the proper width (prior to use) and a drop of ink is placed between them. Do not dip the pen, or place more than one drop of ink between the nibs. Clean the pen point frequently. Hold the pen against the straightedge or curve in an upright position at a steady angle during use.

Sandpaper

Sandpaper may be used to sharpen pencils, to obtain graphite from pencils for smudging with a stump, and to remove irregu-larities from cut surfaces of illustration board.

Sandstone. See Erasers.

Scissors

A good quality scissors is sometimes helpful for use on drawing paper, although a razor or an X-Acto knife is usually preferable.

Scraper board. See Papers.

Scratch board. See Papers; Scratch board tools.

Scratch board tools

Various metal instruments that fit an ordinary pen holder and that are designed to scratch lines and gouges of various widths on scratch board surfaces. Most engraving tools are adaptable to this use, as are needles, nails, and flat gouging instruments. Multiline scratch-ing tools are equipped with rows of teeth that scratch several parallel lines with one stroke.

Sharpeners

A knife, razor blade, or sandpaper is best for obtaining the desired pencil point. Exposing too much lead allows it to break off easily when pressure is used.

Shield. See Erasers.

Smudger. See Stump.

Speedball pen. See Pen points.

Stomp. See Stump.

Stipple board. See Papers.

Strathmore. See Papers.

Stump

A short, thick, pencil-shaped roll of paper, soft leather, or similar material, usually with a fairly blunt point, used for rubbing pencil or charcoal to achieve subtle graduations in tone representing light and shade. The paper stick in candy suckers, when sharpened and rubbed in graphite or pencil smears on sandpape, makes an excellent shading stump. A stump can easily be made by rolling paper tightly and sharpening with a penknife or razor. Sometimes graphite is added to the surface before it is rolled and sharpened. Stumps are also called French stump, stomp, tortillon, or smudger.

Template. See Lettering guide.

Texture board. See also Paper.

A general term for any illustration paper with a pebbly or grainy surface. The texture is brought out by shading.

Tortillon. See Stump.

Tracing Paper. See Papers.

Tracing Box
A device usually made with a frosted glass surface set over fluorescent or other light fixtures so that light may be transmitted through drawing paper placed on top of the glass; also called a copy box or light box. A window can be used for small tracing work in daylight. An old, lift-up typewriter desk can be converted to a tracing table by removing the hinged center typewriter platform and replacing it with glass; two fluorescent tubes placed in the center drawer position supply the light.

Transfer-type lettering. See Lettering.

Transparent acetate. See Acetate sheet overlay.

Transparent overlay. See Acetate sheet overlay.

Triangle
A plastic or metal mechanical drafting tool used for drawing slopes and angular lines. Most commonly used are ones with 30 degree-60 degree-90 degree angles and 45 degree-90 degree-45 degree angles. They are available in many lengths from 4 inches to 18 inches. They are designed to be used with a T square.

T square
A long, straightedged wood or plastic tool equipped with an offset tip mounted at right angles to the edge. It is used for drawing horizontal lines, aligning drawings and legends, and as a guide for triangles and lettering sets.

Two-sided tape. See Adhesives.

Vellum. See Papers.

Wax pencil. See China marking pencil; Lithographic pencil or crayon.

Whetstone. See Carborundum.

Wrico. See Lettering.

X-Acto knife

A brand name of a knife with a metal handle and a chuck (mechanical device) that holds a knife blade. Blades are available in a variety of sizes and shapes at most hobby stores. The No. 11 blade is useful for scratch board cuts and acetate overlays. The maker recommends wearing goggles when using this knife because of the extreme sharpness of the blade.

Zipatone. See also Acetate sheet overlay; Craftint.

A brand name of a type of transparent acetate overlay with a printed pattern on one side and a waxy or adhesive backing for affixing the overlay to a drawing. The term has become general for all types of overlays. Other brands of this type include Artype, Chart-pak, Craftint, Formatt, and Letraset.

Align. To bring into line.

Arc. A curved line.

Burnish. To rub with a tool to polish, smooth, or cause to adhere to another surface.

Critique. To judge; to give an opinion of the merits of a drawing being examined.

Cross-hatch. To mark with two sets of parallel lines that intersect one another.

Diminution. Decreasing in size.

Distortion. A lack of proportional or proper shape in an image or drawing.

Dorsal. Relating to the back side of an animal as distinct from the belly side.

Draft. To draw a preliminary sketch, version, or plan.

Ellipse. A geometric shape that is a closed oval curve.

Engrave. To form figures, letters, or devices by cutting into a surface.

Foreshorten. To draw in perspective so that objects (particularly linear objects) will appear smaller in depth.

Gradation. A gradual passing from one tint or shade to another.

Hachure. One of a series of fine lines used for shading and/or denoting surfaces in relief.

Halftone. A process of making photos or plates for printing illustrations where the lights and shadows are produced by fine dots.

Highlight. A light spot or area produced by more intense light.

Horizontal. Operating in a plane parallel to the horizon or to a base line.

Illuminate. To supply light or brighten an object with light.

Illustration. A picture drawing or diagram that helps make something clear, a visual explanation.

Legend. The explanatory comment accompanying an illustration.

Linecut. A photoengraving from a line drawing.

Medium. Material or technical means of artistic expression.

Mount. To attach to a support for use or display.

Nib. The tip or point of a pen, usually separated by a slit in the metal.

Opaque. A property of not allowing light through.

Parallel. Extending in the same direction, everywhere equidistant.

Perspective. The technique of drawing lines and shadows so as to give an illusion of depth, height, and breadth; the spatial relation of three-dimensional objects as they might appear to the eye in a two-dimensional drawing.

Ply. One of several layers.

Portfolio. An assemblage of representative artwork; also a portable case used for carrying drawings.

Proportion. The relation of one part to another or to the whole with respect to magnitude, quantity, or degree; a proper balance.

Reduction. An illustration made smaller by photographic or meechanical means.

Sketch. A quick, unfinished drawing representing the chief features of an object or scene, often made as a preliminary study.

Stipple. To produce light and shade in graphic art by means of dots; one of the dots used in the process.

Symmetry. Balanced proportions; correspondence in size, shape, and relative position of parts on opposite sides of a body.

Texture. The surface characteristics of a body or substance that impart a characteristic light absorbance or reflectivity.

Tone. The effect of light and shade together; the overall degree of darkness of an object or drawing.

Tract. An area or region as distinct from a larger one, i.e., a white fur tract on a spotted rabbit.

Translucent. A property permitting diffused light through so objects cannot be seen clearly.

Transparent. A property permitting light through so that objects are clearly visible.

Vanishing point. A point at which a group of receding parallel lines seems to meet when represented in linear perspective.

Ventral. Relating to the belly side of an animal as distinct from the back side.

Vertical. Upright; perpendicular to the plane of the horizon or to a primary axis.

SUGGESTED REFERENCES

Cutler, Merritt. 1960. How to Cut Drawings on Scratchboard. New York: Watson-Guptill. 88 pp.

D'Amelio, Joseph. 1964. Perspective Drawing Handbook. New York: Leon Amiel. 96 pp.

Guptill, Arthur L. 1976. Rendering in Pen and Ink. Second ed. Susan Meyer, ed. New York: Watson-Guptill. 255 pp.

Knudsen, Jens W. 1966. "Scientific Illustration." In Biological Techniques. New York: Harper and Row. Pp. 446-489.

Mayer, Ralph. 1970. The Artist's Handbook of Materials and Techniques. 3d rev. ed. New York: Viking Press. 749 pp.

Papp, Charles S. 1976. Manual of Scientific Illustration. 3d ed. Sacramento: American Visual Aid Books. 339 pp.

Perard, Victor. 1957. Drawing Animals. New York: Grosset & Dunlap. 48 pp.

Pitz, Henry C. 1957. Ink Drawing Techniques. New York: Watson-Guptill. 144 pp.

Ridgway, John L. 1938. (1978 reprint) Scientific Illustration. Stanford, Calif.: Stanford Univ. Press. 173 pp.

Simmons, Seymour, and Marc S. A. Winer. 1977. Drawing, The Creative Process. Englewood Cliffs, N.J.: Prentice-Hall. 272 pp.

Snyder, John. 1973. Commercial Artist's Handbook. New York: Watson-Guptill. 264 pp.

Wood, Phyllis. 1979. Scientific Illustration. New York: Van Nostrand Reinhold. 148 pp.

Zweifel, F. W. 1961. Handbook of Biological Illustration. Chicago, Ill.: Phoenix Science Series, University of Chicago Press. 131 pp.

I N D E X